Digital Broadband
Networks & Services

Other Related Titles

Digital Broadband Networks & Services

Bertil C. Lindberg

McGraw-Hill, Inc.
New York   San Francisco   Washington, D.C.   Auckland   Bogotá
Caracas   Lisbon   London   Madrid   Mexico City   Milan
Montreal   New Delhi   San Juan   Singapore
Sydney   Tokyo   Toronto

Library of Congress Cataloging-in-Publication Data

Lindberg, Bertil C.
    Digital broadband networks and services / by Bertil C. Lindberg.
        p.   cm.
    Includes index.
    ISBN 0-07-037936-X
    1. Digital communications.   2. Broadband communication systems.
    I. Title.
    TK5103.7.L56   1994
    384—dc20

94-26706
CIP

1 2 3 4 5 6 7 8 9 0   DOC/DOC   9 0 9 8 7 6 5 4

ISBN 0-07-037936-X

*The sponsoring editor for this book was Stephen S. Chapman, the edit-
ing supervisor was Caroline R. Levine, and the production supervisor
was Suzanne W. Babeuf. This book was set in Century Schoolbook by
McGraw-Hill's Professional Book Group composition unit.*

Printed and bound by R. R. Donnelley & Sons Company.

Contents

Part 3   Digital Broadband Network Technology

To my wife,
children, and grandchild

Preface

This is the first book published on the new evolving field of digital broadband networks and services. These networks are also known as *electronic superhighways* This book will familiarize you with this concept and technology and it is intended as a tutorial. There are some mathematical formulas in the text, but do not let them discourage you. You can skip them without loosing the context of the book. Actually, the mathematical formulas are there for those who prefer that kind of shorthand to the longer text.

Digital broadband technology is still being developed and is so new that the standards bodies have not kept up with it. Thus some important standards have only been available to the author in draft form. They are, however, likely to be adopted in the form discussed in the book. The equipment for broadband networks, in particular switching equipment, is not mature yet. The book discusses most of the different technologies being considered or available. It will assist you in making the right decisions regarding the selection of technology and equipment.

Chapter 1 gives an overview of the entire subject. Some potentional applications for broadband networks are discussed in Chapter 2. As the deployment of such networks matures, users are likely to develop additional and new applications. The basics of digital technology, in particular digitizing, coding, protocols, and packet switching, as well as compression techniques, are discussed in Chapter 3. Compression is an important aspect of broadband technology, since it reduces the requirements for bandwidth. New concepts and buzz words such as asynchronous transfer mode (ATM) and broadband integrated services digital networks (B-ISDN) are introduced in this chapter. They provide the basic technology for the new broadband services.

Part 3 (i.e., Chapters 4 through 11) contains the major emphasis of the book. Subscriber loops (user access links) are discussed in Chapter 4, and user (subscriber) interfaces are discussed in Chapter 5. This includes access links and protocols for packets, frames, and cells, as well as accesses to switched services, such as switched multi-megabit data service (SMDS), broadband integrated services digital networks (B-ISDNs), and asynchronous transfer mode (ATM). Chapter 6 covers network management technology. Local area network (LAN) technology and, particularly, new high-speed LANs are dicussed in Chapter 7. The actual superhighways, the broadband digital trunks and satellite links, are dealt with in Chapters 8 and 9.

The connection and switching of broadband links are treated in Chapters 10 and 11. Packet switching, circuit switching, asynchronous transfer mode (ATM), frame relay, and switched multimegabit data services (SMDS) are discussed from the interconnection point of view. These chapters give you ideas about how to extend and interconnect your current networks into future broadband metropolitan and wide area networks. In Chapter 12, components for broadband technology, in particular very large scale integrated (VLSI) circuits and optical components, are covered.

Part 5 treats the political aspects such as the involvement of foreign countries, the Federal Communications Commission, the U.S. Congress, etc. The balance between regulated and free markets, as well as the competition between conventional telephone companies and cable television companies for a slice of the new interactive broadband market, is discussed. In Part 6, the overall management of broadband networks is covered. This includes a chapter on congestion control, security, and encryption, as well as a chapter on the performance criteria likely to be demanded by users and the quality of service that users can expect and demand.

Acknowledgments

The author extends his appreciation and thanks to the many people among research institutions and manufacturers who contributed to the book. They are so many that they cannot be individually listed. Special thanks go to the anonymous reader(s) of the manuscript. Dr. Joan L. Mitchell read and contributed to the part on compression. All added to the value of the book.

Several companies and publishers gave me permission to use their copyrighted material and I thank them.

Thanks go to the editors and producers at McGraw-Hill, in particular Neil Levine, Steve Chapman, Carol R. Levine, and Suzanne Babeuf.

Bertil C. Lindberg

List of Abbreviations

2BIQ	two binary bits contained in one quaternary symbol
5ESS-2000	an AT&T switch model
AAL	ATM adaptation layer
AC	access class
ac	alternating current
ACK	acknowledgment (of reception)
ACM	address complete message
ACM	Association for Computing Machinery
ACTS	advanced communications technology satellite
AD	access device (Ericsson)
ADM	add-drop multiplexor
ADPCM	adaptive differential pulse code modulation
ADSL	asymmetric digital subscriber line
AIC	ATM input circuit
AIEE	American Institute of Electrical Engineers (now IEEE)
AIN	advanced intelligent network
AM/PM	amplitude modulation/phase modulation
ANet	Asynchronous Network (Northern Telecom)
ANM	answer message
ANSI	American National Standards Institute
AOC	ATM output circuit (Ericsson)
ASCII	American Standard Code for Information Interchange
ASIC	application specific integrated circuit
ASM	ATM switch module (Ericsson)
ASX-1000	an Alcatel switch model
Async.	asynchronous data transfer
ATM	asynchronous transfer mode
AXC-2000	an AT&T switch model
BAP	broadband application part
baud	number of discrete conditions or signal events per second

BCH	Bose-Chaudhuri-Hocquenquem
BCN	backward congestion notification
BER	bit-error-rate
B-ET	broadband exchange termination
BICMOS	combined bipolar and CMOS semiconductor
B-ISDN	broadband integrated services digital network
Bisync.	bisynchronous data transfer (IBM)
bit/s	binary digit per second
BIU	bus interface unit
BLER	block error rate
BLSR	bidirectional line switched ring
BNS-2000	an AT&T switch model
B-NT	broadband network termination
BORSCHT	battery, over-voltage, ringing, supervision, codec/control, hybrid and testing
BORCHT	battery, over-voltage, ringing, supervision, hybrid and testing
B-TA	broadband terminal adapter
B-TE	broadband terminal
byte	8 binary digits or bits. *See also* octal
CAD	computer-aided design
CAM	computer-aided manufacturing
CAP	carrierless amplitude and phase
CAT	computerized axial tomography
CATV	community antenna television
CBDS	connectionless broadband data service
CBI	cell bearer interface (Ericsson)
CBR	constant bit rate
CCIR	Comité Consultatif International des Radiocommunications; International Radio Consultative Committee
CCIS	common-channel interoffice signaling
CCITT	Comité Consultatif International Télégraphique et Téléphonique; International Telegraph and Telephone Consultative Committee
CCS	common-channel signaling
CD	compact disc
CDMA	code division multiple access
CD-ROM	compact disc read-only memory
CELP	code-excited linear prediction
CIF	common intermediate format
CLIP	calling line identity presentation
CLP	cell loss priority
CMOS	complementary metal-oxide semiconductor

CMTT	Commission Mixte des Transmission Télévisuelles et Sonores
codec	coder-decoder
CPCS	common part convergence sublayer
CPE	customer premises equipment
CPU	central processing unit
CS	convergence sublayer
CSA	carrier serving area
CSMA/CD	carrier sensing multiple access with collision detection
CSU	channel service unit
D/A	digital-to-analog
dB	decibel
dc	direct current
DCE	data circuit-termination equipment
DCS	digital cross connect system
DCT	discrete cosine transform
DES	Data Encryption Standard
DM	delta modulation
DMS	digital multiplex switch (Northern Telecom)
DMT	discrete multitone
DPCM	differential pulse code modulation
DPN-100	a Northern Telecom switch model
DQDB	distributed queue dual bus
DS-n	digital signal level n
DSC	distributed source control
DSP	digital signal processing
DSS1	digital subscriber signaling number 1
DSS2	digital subscriber signaling number 2
DSU	data service unit
DTE	data terminal equipment
DUP	data user part
DXI	data exchange interface
E-n	European digital signal level n
ECL	emitter-coupled logic
ECN	explicit congestion notificatio
ECSA	Exchange Carriers Standards Association
EFS	error-free second
EIA	Electronic Industries Association
EIM	external interface module (Northern Telecom)
ENet	Enhanced Network (Northern Telecom)
EOC	embedded operations channel
EOL	end of line
ESTI	European Telecommunications Standards Institute
ET	exchange termination

ETT	external protocol (or transmission) termination (Alcatel)
EWSM	Elektronisches Wähler System, M (Siemens)
FC	fiber channel
FCC	Federal Communications Commission
FCI	forward congestion indicator
FDCT	forward discrete cosine transform
FDDI	fiber distributed data interface
FETEX	Fujitsu electronic telecommunication exchange
FFOL	FDDI follow-on LAN
FFT	fast Fourier transform
FIFO	first-in, first-out
FIL	filter
FITL	fiber-in-the-loop
FNS	Fujitsu Network Switching of America
FOMAU	fiberoptic media access unit
FSK	frequency shift keying
FTTC	fiber-to-the-curb
FTTH	fiber-to-the-home
GaAs	gallium arsenide
Gbd	gigabaud (= 1,000,000,000 bauds)
Gbit	gigabit (= 1,000,000,000 binary digits)
Gbit/s	gigabit per second (= 1,000,000,000 binary digits per second)
GCNS-2000	an AT&T switch model
GHz	gigahertz (= 1,000,000,000 periods per second)
h	hour
Hn	CCITT channel standard type n
HDLC	high-level data link control
HDSL	high-bit-rate digital subscriber line
HDTV	high-definition television
HEC	header error check
hi-fi	high fidelity
HIPPI	high-performance parallel interface
Hz	hertz (= period per second)
IAM	initial address message
IC	integrated circuit
ICC	interface conversion chip (Alcatel)
ICC	internodal communications controller (Northern Telecom)
IDCT	inverse discrete cosine transform
IDDD	international direct distance dialing
IDU	interface data unit

IEC	International Electrotechnical Commission
IEEE	The Institute of Electrical and Electronic Engineers, Inc.
IFFT	inverse fast Fourier transform
IFRB	International Frequency Registration Board
IN	intelligent network
IP	Internet protocol
IRE	Institute of Radio Engineers (now IEEE)
IS	international standard
ISCP	ISDN service control part
ISDN	integrated services digital network
ISDN-UP	integrated services digital network user part
ISE	integrated switching element (Alcatel)
ISO	International Organization for Standardization
ITSEC	Information Technology Security Evaluation Criteria
ITU	International Telecommunication Union
ITU-T	Telecommunications Standardization Sector of the International Telecommunication Union
IXC	interexchange carrier
JBIG	Joint Bi-level Image Experts Group
kbit/s	kilobit per second (= 1000 binary digits per second)
kHz	kilohertz (= 1000 periods per second)
km	kilometers (= 1000 meters, $\sim$ 0.62 miles)
L2-PDU	level 2 protocol data unit
LAN	local area network
LAP	link access procedure
LAPB	balanced link access protocol
LAPD	link access procedure/protocol on the D channel
LATA	local access and transportation area
LATM	local ATM
LCF-PMD	low-cost fiber media-dependent physical layer
LEC	local exchange carrier
LED	light-emitting diode
LEO	low earth orbit (satellites)
LLC	logical link control
LPC	linear predictive coding
LPS	less probable symbol
m	meter (= 1.0936 yards)
m	milli (= 1/1000)
m	minute
MAC	medium access control
MAN	metropolitan area network
Mbd	megabaud (= 1,000,000 bauds)

Mbit	megabit (= 1,000,000 binary digits)
Mbit/s	megabit per second (= 1,000,000 binary digits per second)
MFJ	modified final judgment
MH	modified Huffman
MHz	megahertz (= 1,000,000 periods per second)
MID	message identifier
MMF	multimode fiber
MMF-PMD	multimode fiber media–dependent physical layer
modem	modulator/demodulator
MOS	metal-oxide semiconductor
MPEG	Moving Picture Experts Group
MPS	more probable symbol
MPSR	multipath self-routing
MR	modified READ
MSC	multislot cell (Alcatel)
MSVC	meta-signaling virtual channel
MTP	message transfer part
MTR	maintenance and test routiner (Alcatel)
MUX	multiplexor
NAK	negative acknowledgment (of reception)
NCP	network control point
NEAX	a NEC switch model
NEXT	near-end crosstalk
N-ISDN	narrowband integrated services digital network
nm	nanometer (= 1/1,000,000,000 m)
NNI	network-node interface
NSA	(U.S.) National Security Agency
NT	network termination
NT1	network termination type 1
NT2	network termination type 2
NTC	network time constant
OAM	operations, administration, and maintenance
OAM&P	operations, administration, maintenance, and provisioning
OBC	on-board controller (Alcatel)
OC	optical carrier
OC-n	optical carrier level n
octal	= 8 binary digits or bits. *See also* byte.
octet	= 8 binary digits or bits. *See also* byte.
OH$^-$	hydroxyl ion
OMAP	operation, maintenance, and administration part
OSI	open systems interconnection
PABX	private automatic branch exchange

PAD	packet assembler/disassembler
PAM	pulse amplitude modulation
PA/R	peak-to-average ratio
PBA	printed board assembly
PBX	private branch exchange
PC	personal computer
PCM	pulse code modulation
PCN	personal communications network
PDH	plesiochronous digital hierarchy
PDU	protocol data unit
pel	picture element
PLCP	physical layer convergence procedure
POP	point of presence
POT	plain old telephony
POTS	plain old telephone service
PSPDN	packet switched public data network
PTI	payload type identifier
PTT	Post Telegraph and Telephone
QAM	quadrature amplitude modulation
QCIF	quarter-common intermediate format
QOS	quality of service
QPSK	quarternary phase-shift keying
RAM	random access memory
RBOC	regional Bell operating company
READ	relative element address designate
REL	release message
RISC	reduced-instruction-set computer
RIU	ring interface unit
RLC	release complete message
RSA	R. Rivest, A. Shamir, and L. Adleman
s	second
SAPI	service access point identifier
SAR	segmentation and reassembly
SBIC	shared bus interface controller
SBM	shared buffer memory (Alcatel)
SC	switch core (Ericsson)
SCCP	signaling connection control part
SDH	synchronous digital hierarchy
SDLC	synchronous data link control (IBM)
S/DMS	SONET/digital multiplex switch
SDU	service data unit
SEAL	simple and efficient AAL
SEP	signaling end point
Si_3N_4	silicon nitride

SiO_2	silicon oxide
SIP	SMDS interface protocol
SLC	subscriber loop carrier
SMDS	switched multimegabit data service
SMX-6000	A Fujitsu switch model
SNA	systems network architecture (IBM)
SNet	SONET Network (Northern Telecom)
SONET	synchronous optical network
SP	switch port (Ericsson)
SPC	stored program-controlled
SRT	self-routing tag (Alcatel)
SS6	signaling system number 6
SS7	signaling system number 7
SSCS	service specific convergence sublayer
SSET	switching system exchange termination
SSP	service switching point
STM	synchronous transport module
STM	synchronous transfer mode
STP	signaling transfer point
STS	space-time-space
STS	synchronous transport signal
SVC	signaling virtual channel
Tn	digital carrier, type n
T1E1	A standards working group of ECSA
TA	terminal adapter
TAT	transatlantic telephone (cable)
Tbit	terabit ($= 1,000,000,000,000$ binary digits)
Tbit/s	terabits per second ($= 1,000,000,000,000$ binary digits or bits per second)
TC	transactional capabilities
TCAP	transaction capabilities application part
TCI	Telecommunications, Inc.
TCSEC	Trusted Computer Security Evaluation Criteria
TDM	time-division multiplex
TDMA	time-division multiple access
TEI	terminal endpoint identifier
TMC	transfer mode converter/switch port termination (Alcatel)
TMN	telecommunication management network
TST	time-space-time
TSU	traffic switching unit (Alcatel)
TTC	Telecommunications Technology Committee (Japan)
TUP	telephone user part
TV	television

UN	United Nations
UNI	user-network interface
UPSR	unidirectional path switching ring
USAT	ultrasmall aperture terminal
UUCP	UNIX-to-UNIX copy
UUS	user-to-user
VBR	variable bit rate
VC	virtual channel
VCI	virtual channel identifier
VF	voice frequency (300 to 3400 Hz)
VHADSL	very high-bit-rate asymmetric digital subscriber line
VHDSL	very-high-bit-rate digital subscriber line
VLSI	very large scale integrated (circuits)
VME	a computer bus
VP	virtual path
VPCI	virtual path connection identifier
VPI	virtual path identifier
VSAT	very small aperture terminals
WAN	wide area network
WDM	wavelength division multiplexing
X.25	a CCITT recommendation
μm	micrometer or micron (= 1/1 000 000 m)
μs	microsecond (= 1/1 000 000 s)

Introduction and Basic Technology

Part

1

Introduction and
Basic Technology

1

Overview

1.1 History

The rapidly increasing demand for communications is driving the telecommunications industry to the use of ever-increasing bandwidths. Simultaneously, digital systems are taking over from analog systems. The result is digital broadband networks and services.

Telecommunication, i.e., communication at a distance, takes many forms. It originated with runners that carried remembered or written messages between persons. Semaphores came into general use during the Napoleonic wars. They were placed on high hills and towers within visual distance of each other, had arms the position of which indicated a letter or number, and permitted the relaying of information from one tower to the next along a communications line. The next step was Morse's telegraph and Bell's telephone.*

The first applications of the telegraph and the telephone were for point-to-point communication. Later, switches were added to allow for the connection of one user to any one of several users in a network.

---

*Actually, the question of whether Antonio Meucci, an Italian-American who spent most of his life on Staten Island in New York, invented and operated a working telephone when Alexander Graham Bell was 2 years old has resurfaced recently in an article by Professor Pier L. Bargellini (former professor at the University of Pennsylvania), entitled, "Books on Meucci probe role in telephone," in *The Institute,* vol. 14, no. 4, April 1990, p. 8 (published by IEEE, the Institute of Electrical and Electronic Engineers). Several books (see Giovanni E. Schiavo, *Antonio Meucci—Inventor of the Telephone,* New York, The Vigo Press, 1958; and Marco Nese and Francesco Nicotra, *Antonio Meucci, 1808–1889,* New York, Italian American Multimedia Corporation, 150 Fifth Avenue, Suite 423) have been written on the subject and are available in the library of the Garibaldi Meucci Museum, 420 Tompkins Avenue, Staten Island, New York, tel. 718-442-1608. The museum also displays replicas of Meucci's telephones.

Generally, we can state that the purpose of telecommunications networks is to provide transmission channels between users.

Within a few years we had local, area, city, regional, national, and international telegraph and telephone networks. Local networks could be found in a single house, factory, etc. Before there were city-wide networks, we had networks that covered an area of a city. From the beginning we found severe competition between forces supporting independent local networks and those supporting national networks.

In most countries, the telegraph and telephone services eventually became monopolies run by the state. Because the telegraph was seen as a speedier version of mail—and later the telephone was seen as an improvement of the telegraph—the same agency, typically called *Post Telegraph and Telephone* (PTT), ran these services. In the United States and many other countries, the telegraph and telephone services remained in the hands of private companies.

Until the 1940s, telecommunications networks consisted of 4000-Hz-wide analog voice/telephone channels and narrowband telegraph channels. The multiplexing of channels onto a broadband carrier system was started in 1918. Broadband analog carrier systems with a bandwidth of about 390 kHz were introduced in 1940. By the 1960s, the top bandwidth had been increased to over 50 MHz.

Digital multiplex systems were introduced in 1962. Each voice-grade channel occupied 64 kbit/s. This norm is still valid today. Today, broadband digital channels with over 1 Gbit/s are in use.

1.2   Types of Networks

Telecommunications links and networks can be classified into several different types. The first links were of the network *point-to-point type* such as the one between Alexander Graham Bell and Thomas A. Watson on June 2, 1875. The first telegraph links also were of the point-to-point type. This type is shown at the top of Fig. 1.1.

A demand for connecting more than one receiver and transmitter to the same circuit soon developed. In the earliest versions, all transmitters and receivers were connected continuously and could operate simultaneously. They formed networks of the *multidrop type,* shown as the second from the top in Fig. 1.1.

In a following step, switches were introduced, switches that established connections between two terminals at a time. They came to be called *circuit-switched networks*. The terminals are usually connected to a switch in the form of a star. In large networks, there are direct links between all major switches, forming a mesh network. These kinds of networks are shown at the bottom of Fig. 1.1

In one version of a telegraph network, all telegraph messages are sent to a central unit for temporary storage. These units, which are

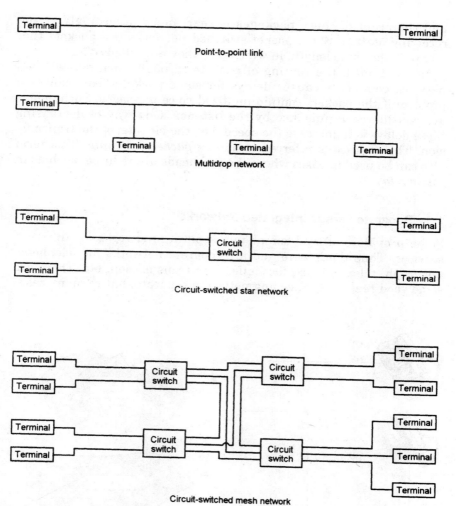

Figure 1.1   Types of networks.

also called *switches,* in turn retransmit the messages to the receivers depending on the address contained in the heading of the message. The entire process is called *store and forward message switching.*

Packet switching is a refinement of store and forward message switching. A message, typically a message with computer data—however, it also can be voice or represent a picture—is divided into packets. Each packet is preceded by an address. Based on this address the packet is transported through a network until it reaches its destination. It is possible that the packets arrive in an order that differs from the order in which they were sent. Thus each packet is numbered, and the order of these packages is rearranged accordingly at the receiving station. Some packet-switched networks use packets that all are of the same length, in which case they are called *cells*.

As mentioned, the routing of packets through a packet-switched network commonly causes delays for some packets. This causes a problem if the packets contain digitized voice messages. Such delays are considered a nuisance by the listener. One way of decreasing these delays is to increase the speed, i.e., the bit rate, of the transmission. This feature is referred to as *fast packet switching*. This term also can be used in cases where the overheads are reduced, such as in *frame relay*.

1.3   Separate versus Integrated Networks

Some providers offered completely separate networks for different services. Thus there were and are separate networks for telephone, telegraph, telex, teletex, facsimile, data transmission, etc. There are some good reasons for keeping networks separate, but in many cases

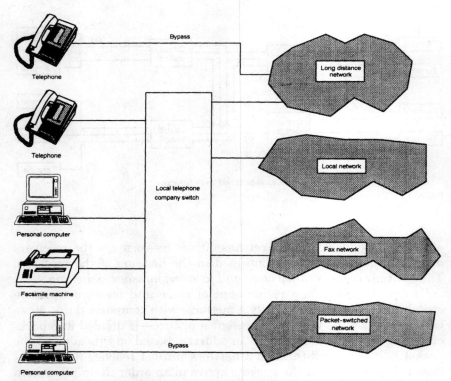

Figure 1.2   Interfacing to different networks through the local telephone company.

it would be less expensive and more efficient to use the same facilities for different services. One reason for keeping the networks separate is that the requirements for bandwidth vary. Thus telegraph and even telex traffic can be handled over channels with a fraction of the bandwidth required for voice or data. On the other side, some types of fast facsimile transmission require wider channels.

The separate networks can be accessed through the local telephone network by means of dial-up, as shown in Fig. 1.2. They also can be reached through private lines that bypass the local telephone companies' networks.

An *integrated network* combines several types of separate networks into one. Typically, its access lines are digital. A variety of different types of terminals can be connected to such a network, as shown in Fig. 1.3.

Certain transmission and switching facilities in a network can be used to create a *virtual network,* as shown in Fig. 1.4. One type of virtual network creates a dedicated point-to-point type of channel between two user terminals. If such a channel passes through a switch, it is permanently set to connect two channels to the user interface, one on each side of the switch. Another type of virtual network can be used to create a private network within a public network. In this case, switches in the network are used to set up and switch connections through the virtual network under control of the users.

In the future, data and video transmissions will require wider bandwidths than those available today. This book discusses the meeting of this demand through wideband transmission channels and channels with flexible bandwidths.

Currently in the United States, separate networks are found for voice, telex, data, and video. We have the ubiquitous and international telephone network for voice. Most data traffic is handled by separate packet-switched data networks, including local area networks (LANs). Most telex traffic is carried on separate networks. We say *most,* because some data and telex traffic is handled over the common telephone network. This is especially true for accessing the special networks. Technically, common telephone facilities are used for data and telex networks; they are just operationally disjoint. The only major video networks existing are so-called cable networks. The original name is *community antenna television* (CATV) *network,* and such a network is used to pick up broadcast television transmission by a central antenna and distribute the same signal over a cable network. These cable networks are generally one-way networks for TV broadcast signals. Only in a few such networks can a user send signals back to the head end of the network, and the bandwidth of these channels is usually just a fraction of the video signal.

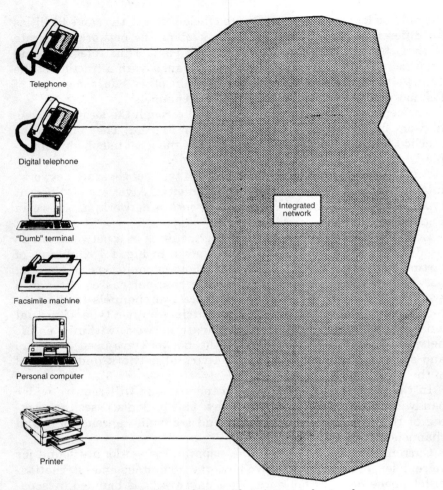

Figure 1.3   Different terminals interfacing with an integrated network.

In February of 1991, the Federal Communications Commission (FCC) granted experimental licenses to several cable television companies to provide cellular radio service, in particular so-called personal communications, using their cable networks as feeders to radio base stations.

1.4   New Name for CCITT

The International Telecommunication Union (ITU), a specialized agency of the United Nations, is an international organization for the

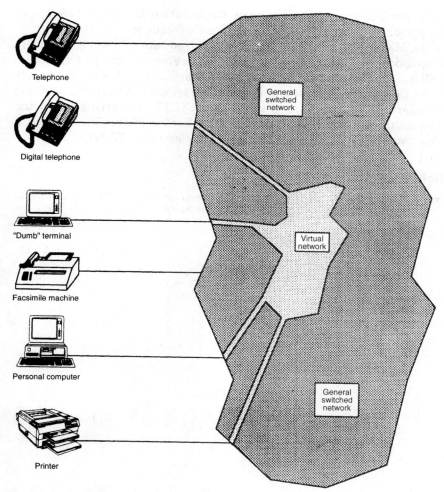

Figure 1.4 Some terminals interfacing with a virtual network.

standardization, regulation, and development of telecommunications. Since 1956, the standardization body within ITU has been known as the *International Telegraph and Telephone Consultative Committee.* It is better known by its acronym, CCITT, which comes from its name in French, *Comité Consultatif International Télégraphique et Téléphonique.* A sister organization, the International Radio Consultative Committee, or CCIR (from *Comité Consultatif International des Radiocommunications*) was responsible for standardization in the field of radio communications. A third organization, the International Frequency Registration Board, or IFRB, was responsible for regulating the use of radio frequencies.

On March 1, 1993, ITU was reorganized, and three sectors were created: the Radio Communications Sector (formerly CCIR and IFRB), the Telecommunications Standardization Sector (formerly CCITT), and the Telecommunications Development Sector. Today the standardization functions of the former CCITT are carried out by the Telecommunications Standardization Sector of the International Telecommunication Union (ITU), with the acronym ITU-T.

Reference

Irmer, Theodor (1994), "Shaping Future Telecommunications: The Challenge of Global Standardization, *IEEE Communications Magazine,* vol. 32, no. 1, January, pp. 20–28.

Chapter

2

Applications

2.1 Introduction

A broadband network can be compared to a superhighway that carries traffic from and to many different locations. Without traffic feed-
ing into the superhighway and traffic leaving it for some destination,
there would be no traffic on the highway. Similarly, there will be no
traffic on a broadband telecommunications network unless there are
sources and sinks of traffic.

Some activities that can generate traffic of high bit rates are listed
in Table 2.1. Often, when people talk about digital broadband networks, they think about two supercomputers that exchange a lot of

TABLE 2.1 Traffic with High Bit Rates

1. Multimedia activities
2. Videoconferencing
3. Video-on-demand
4. Desktop publishing
5. Computer-aided design and manufacturing (CAD/CAM)
6. Medical imaging
7. Color facsimile
8. High-definition television and video
9. High-speed file transfer
10. Supercomputing
11. Distance learning
12. Molecular modeling
13. Financial modeling
14. Virtual reality
15. Engineering visualization
16. Geophysical modeling
17. Animation
18. Collaborative design

TABLE 2.2   Types of Terminals Generating High Bit Rates

1. Computers, from personal computers to supercomputers
2. Data terminals
3. Digital facsimile machines
4. Information data banks
5. Real-time television broadcasts
6. Scanners of medical images (CAT, x-ray, etc.)
7. Servers of information
8. Still-picture data banks
9. Video cameras, including remote video cameras
10. Video data banks (videotapes)
11. Videophones
12. Videoconferencing terminals
13. Video terminals
14. Workstations

traffic over a point-to-point channel. There are, however, more sources than supercomputers that feed traffic to broadband networks, and even if most traffic is point to point, the individual "points" vary from transmission to transmission. We are really talking about a *network*.

Some terminals that generate high bit rates, i.e., broadband traffic, are listed in Table 2.2. Some types of terminals generate the same amount of traffic over time; others occasionally generate bursts of traffic at high bit rates. The traffic from a number of terminals can be multiplexed and made to enter a network at a relatively high bit rate. For example, the external traffic from terminals on a broadband LAN is frequency multiplexed and offers a relatively high bit rate. The more common baseband LANs use a form of time multiplexing. Figure 2.1 shows examples of these kinds of traffic patterns.

Two factors reduce the amount of bandwidth required. One is compression, and the other is the fact that transmission can be spread out in time. Unless instant transmission is required, as in the case of real-time voice and video transmissions, bandwidth can be traded for delays. Thus a 50-Mbit file can be transmitted in a little over 1 s over a 45-Mbit/s link or in 13 min over a 64-kbit/s link. Over an analog link with modems having a speed of 4800 bit/s, the transmission would take 2 h 53 min. In other words, a broadband network saves a lot of time. Table 2.3 shows some examples.

2.2 Video-on-Demand

Eventually, the largest demand for broadband services will be generated by *video-on-demand*. This refers to a feature where a potential viewer requests a certain sequence of video and that sequence is

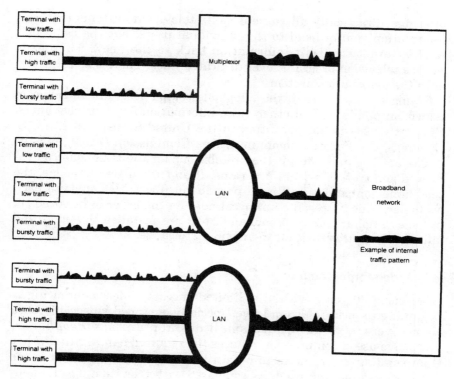

Figure 2.1 Traffic patterns.

instantly transmitted to the user. The sequence can be a movie, a television show, a video sequence that is part of a multimedia presentation, a lecture, or any other type of video presentation. The video sequences can be available from central or distributed video banks, or they can be live presentations including videoconferences. Most cable television companies have the capability to offer this kind of service today, e.g., in the form of pay-TV. Some cable companies are installing systems with some 500 channels, enough for a variety of

TABLE 2.3 Duration of Transmissions

Item	Bits/item	Transmission speed				
		4800 bit/s	64 kbit/s	1.5 Mbit/s	44 Mbit/s	600 Mbit/s
An x-ray	580 Mbit	33 h 34 min	2 h 31 min	6 min 27 s	13 s	0.9 s
An encyclopedia	280 Mbit	16 h 12 min	2 h 31 min	3 min 7 s	6.4 s	450 ms
A novel	10 Mbit	34 min 43 s	2 min 36 s	6.7 s	227 ms	16 ms
A facsimile page	250 kbit	52 s	4 s	0.17 s	6 ms	0.4 ms

offerings. Practically all current cable television systems transmit information from a head to the individual users, but the cable networks cannot transmit information back to the head. The heads receive television programs over dedicated lines. There are no switches in the distribution system.

Seeing a large market, the telephone companies want to generate television programs and other video information for their customers. Currently, this is not permitted in the United States, but the U.S. Congress, the Federal Communications Commission (FCC), and the courts are beginning to lift the prohibitions for the telephone companies to generate television programs. If and when this happens, the telephone companies will need broadband fiberoptic cables to each customer. This represents an investment on the order of hundreds of billions of dollars. Once broadband links are installed to homes and offices, many other uses for them will materialize.

2.3 Videoconferencing

After some 30 years with very limited success, videoconferencing is beginning to pick up. From large, specially equipped videoconference rooms we are going toward small desktop units, *videophones*. Extensive use of compression reduces the required bandwidth. Rather than sending 30 frames/s to obtain television quality, much lower frame rates are often used. As compared with television, not to mention high-definition TV, users may accept much lower resolutions or reduction of the size of the display window to a fraction of a full TV screen or computer monitor window. The result is that pictures acceptable to some people can be sent over 4-kHz analog voice lines. In other words, broadband links are not required. However, the quality of the video transmissions in resolution and frame rate will increase with higher transmission bandwidths. A bandwidth of 1.5 Mbit/s can accommodate black-and-white video transmissions with a size of 320×240 picture elements (pels) (for one-quarter window) at 15 frames/s using 8 bit/pel and a compression ratio of 1:9, as shown in Table 2.4.

TABLE 2.4   Required Bandwidth for Video Transmission at a Compression Ratio of 1:9

Window resolution	30 frames/s	15 frames/s
700×525 pels and 24 bit/pel	29.4 Mbit/s	14.7 Mbit/s
640×480 pels and 8 bit/pel	8.2 Mbit/s	4.1 Mbit/s
320×240 pels and 8 bit/pel	2.05 Mbit/s	1.02 Mbit/s

Bandwidth is left over for audio, etc. The table also shows some other window sizes and frame rates as examples, including that of commercial color television transmission. In other words, broadband links may only be required for the transmission of high-quality color video.

2.4 Medical Imaging

The possibility for a medical doctor far away from a patent to look at x-rays, CAT scans, and other medical images will enhance the profession. Small medical clinics without radiologists will be able to send images to a specialist instantaneously for his or her review and opinion. Doctors at different locations will be able to see and discuss the same image.

The transmission of medical images requires large bandwidths for several reasons. To be meaningful, high-resolution images are required. Further, the medical profession does not accept image compression that might discard important information. A dot might indicate the beginning of a cancer.

Besides the transmission of images, entire medical files may need to be transmitted without delay. The state of North Carolina is building a broadband network for this purpose, for distance learning, as well as for state administrative purposes.

2.5 Distance Learning

The current trend is to have the classroom come to the student rather than having the students come to the classroom. This increases the efficiency of the teachers and reduces the students' traveling time. The students learn at a distance. A student will have the option to use a canned lecture at his or her leisure or to participate in a live lecture, a kind of conference call.

This kind of teaching and learning requires transmission of voice, images, and video. While canned lectures can be disseminated on CD-ROMs, etc., live lectures require broadband links. Several governmental and private entities are exploring the building of broadband networks for distance learning and other applications.

2.6 Home Shopping and Banking

Home shopping implies that a user can sit in his or her home or office, with catalogs and promotion material, and place orders for merchandise and services. Similarly, *home banking* refers to a situation where a user can transfer funds, obtain statements, and conduct other bank business from a personal computer. Obviously, the computer cannot yet dispense cash or accept deposits.

Simple cases of home shopping and banking can be handled by transmitting low-speed text files and thus do not require broadband facilities. However, we can expect vendors to want to display pictures, including color ones, as well as video on the screen of the potential customer. In such cases, broadband will be required.

2.7   Computer-Aided Design and Manufacturing (CAD/CAM)

Computer-aided design (CAD) and computer-aided manufacturing (CAM), as well as the combination of the two, computer-integrated manufacturing (CIM), involve the transmission of images between terminals. Some of the images are three-dimensional and require a high transmission bandwidth.

In *collaborative design*, a team of designers, dispersed by location, work on the same design. In order to collaborate, they need to communicate. Depending on the type of design, they may need a significant amount of bandwidth to transmit designs between them.

2.8   Engineering Visualization

Besides transmitting three-dimensional pictures of devices, the trend is toward programs that let you visualize the turning of a device in order to look at it from different angles and in different environments.

2.9   Virtual Reality

Virtual reality, which started as a feature of simulation systems for the training of aircraft pilots, is spreading. It is part of many electronic games. From "boxes," it will develop into on-line interactive systems.

2.10   Modeling

Computer programs for modeling of many situations will require transmission over broadband links. This includes geophysical, molecular, and financial modeling. The first two involve three-dimensional models. Financial modeling is made in many dimensions, beyond three. The transmission of these models between terminals will require large bandwidths.

2.11   Animation

Animation, i.e., a sequence of two-dimensional (or even three-dimensional) pictures, will appear in business and residential environments. It will be used in advertising, teaching, and modeling. A relatively large amount of bandwidth is required to transmit animations.

2.12   Information Data Banks

In the future we will find a significant number of data banks that offer text, pictures, and video. Transmitting this information from the data bank to the terminals of users will require a significant amount of bandwidth.

2.13   Multimedia Activities

Users want to talk and videoconference with each other, as well as exchange data, pictures, and videos, more or less simultaneously. On conventional networks, the user would have to use a separate link for each of these services. The advantage with future systems is that all these *multimedia* activities can be performed on a single link to the network. Future broadband links will permit a mixture of services on the same physical link. This is one of the major advantages of the new network systems.

2.14   Desktop Publishing

Desktop publishing involves the creation of text and images, adding gray scale or color pictures, "discussing" these things with others, and sending the final product to the printer/publisher. From the broadband transmission point of view, it is significant that text and images are transmitted between the same terminals (author, source, reviewer, and publisher). A broadband integrated services digital network (B-ISDN) makes it possible to transmit all this information simultaneously over the same link.

2.15   Other File Transfers at High Bit Rates

It is impossible to list all future applications that will require transmissions over broadband links and networks. The imagination of humans is the limit!

Basic Technology

3

Digital Technology

3.1 Introduction

As stated earlier, this book describes data and telecommunications networks that transport information in digital format at speeds ranging from millions of bits per second to billions and beyond. Technologies, applications, and operations are discussed.

The basic digital technology is covered in this chapter. The digitizing and coding of information are discussed, as are communications protocols and data compression techniques. The compression of digital information is an important means of saving bandwidth. Theoretically, digitized color television pictures (with today's normal resolution) require a bandwidth of several million bits per second. At the same time, such video transmission can be compressed to 56,000 bit/s for videoconferencing purposes—a tremendous savings in bandwidth. Even with the "unlimited" bandwidths of fiberoptic links, all the digital traffic anticipated could not be transported economically without signal compression.

3.2 Analog versus Digital Networks

In telecommunications, analog signals consist of continuously varying electrical quantities, while digital signals consist of discrete elements, typically in the form of pulses. The first telecommunications networks, the telegraph networks, used digital signals such as the Morse code. In some cases, *digital* is synonymous with *binary,* but a tristate code such as Morse code (long, short, absent) fits some definitions of digital.

Telephone signals are inherently analog; i.e., the electric signal varies with the level of the sound. In the early 1960s, Bell

Laboratories found that the quality of the transmission of sound signals between switches could be improved and accomplished at a lower price by digitizing the signals. Out of this came the T1 system in North America. This digital system allows for the simultaneous transmission of 24 digitized voice-grade signals on two pairs of twisted copper wires, one pair for each direction. The analog signal is sampled at a rate of 8000 per second. Theoretically, this allows for the transmission and reconstruction of analog signals with a frequency of up to 4000 Hz. This is just above the upper limit of most analog telephone channels, which is 3400 Hz. Sometimes out-of-band signaling and data-over-voice reach to 3700 Hz. The sampling rate of 8000 per second has become a worldwide standard, even though Rolm used 12,000 in its digital PBX switches for a while. Each sample is turned into a digital code that typically consists of 8 bits. These 8 bits represent the amplitude of the analog signal at the time of the sample. The 24 channels of a T1 carrier and frame bits are multiplexed and interwoven to a single bit stream with a rate of 1.544 Mbit/s [8000 × (8 × 24 + 1) = 1,544,000]. Older T1 systems use in-band signaling, and certain bits are "borrowed" for this purpose. This means that only 56,000 bit/s rather than 64,000 bit/s is available all the time. This does not matter to voice but is detrimental for data transmissions. The latest versions of T1 systems offer "clear signals" with a bandwidth of 64 kbit/s.

In Europe, another system for multiplexing digital channels is used, one based on CCITT Recommendation G.702. The major difference is that two separate channels are used for framing and signaling and that the voice channels are free of signaling. This means that a clear signal of 64,000 bit/s is available. Thirty such channels and two signaling channels are multiplexed to form a single channel with a bit rate of 2.048 Mbit/s.

These 1.544- and 2.048-Mbit/s channels are, in turn, multiplexed into channels with higher bit rates. Table 3.1 shows the digital signal

TABLE 3.1   North American Digital Signal Level Hierarchy

Digital signal level	Bit rate (kbit/s)	Equivalent number of (64 kbit/s) VF channels
DS-0	64	1
DS-1	1,544	24
DS-1C	3,152	48
DS-2	6,312	96
DS-3	44,378	672

SOURCE: Compiled from AT&T data.

TABLE 3.2   CCITT Digital Signal Level Hierarchy

Digital signal level	Bit rate (kbit/s)	Equivalent number of (64 kbit/s) VF channels
1st	2,048	30 + 2
2d	8,448	120 + 8
3d	34,368	480 + 32
4th	139,264	1920 + 128
5th	Approx. 565,000	7680 + 512

SOURCE: CCITT Recommendation G.702.

levels and bit rates used in North America. Digital signal levels DS-0, DS-1, and DS-3 are the most common today. Table 3.2 shows the CCITT hierarchy as per Recommendation G.702. In Europe, the G.702 digital signal levels are called E-1 through E-5.

The demand for data communications raised the question of whether these digital channels could be used directly for transporting data. There are several pros as well as problems involved. Once digital transmission channels are available, it seems advantageous to use them for data transmission. The alternative employed in analog networks is to use modems to convert the digital signals into analog signals, which can be transported over an analog telephone network.

One problem is that the North American T1 system is designed for voice transmission only. Digitized voice is rather insensitive to missing bits and thus allows for a rather high bit error rate, a much higher rate than data can sustain. Thus the digital transmission lines have to be reconditioned in order to carry data. Another but more expensive way is to use error-correcting codes.

The specific coding systems used in the older North American T1 carrier systems also cause problems when these systems are used for data transmission. Thus they lack transparency due to the use of a bipolar line code and the in-band signaling that "robs" bits from the data stream. The new signaling system 7 (SS7) will be used instead of in-band signaling in the future.

Further, it is not enough to transport signals in digital form between terminals and switches; the switches themselves have to be able to transport digital signals through them. This is definitely not true for most analog switches. A so-called electronic switch does not necessarily transport signals in digital form. Some older electronic switches use computer technology to establish connections through the switch, connections that are basically analog.

The world trend is to convert analog transmission lines and analog switches to digital networks.

3.3   Digitizing and Coding

The Institute of Electrical and Electronic Engineers (IEEE) defines *coding* as "a process of transforming messages or signals in accordance with a definite set of rules" (IEEE, 1977). Digitizing is a form of coding. Other forms will be discussed later in this book.

Information that takes the form of a *yes* or *no* is in digital form. A yes or a no also can be represented by a *one* (1) or a *zero* (0). Other forms of digital representation are a pulse or a lack of a pulse.

Any analog signal can be converted to a digital signal through sampling and quantizing. *Sampling* means that a sample of the analog signal is taken at periodic intervals, as shown in Fig. 3.1. When digitizing voice signals in the telephone industry, samples are typically taken 8000 times per second. At each sampling period, a pulse whose height corresponds to the amplitude of the analog signal is created. This process is called *pulse amplitude modulation* (PAM). Each PAM pulse is then converted into a binary number representing the amplitude, called *quantizing.*

In the *quantizing,* each sample amplitude is compared against a continuous list of discrete sample intervals (typically 256) and allocated to the proper one. Eight binary digits are typically used in quantizing voice, resulting in 256 sample intervals ($2^8 = 256$). Each discrete interval is described by a code, a binary number. The entire process is called *pulse code modulation* (PCM).

The CCITT recognizes two different types of *companding* (compressing/expanding) used in quantizing, A-law and μ-law. The μ-law is used in North America and Japan, while the A-law is used in the rest of the world. The normalized expression of the μ-law is

$$y = \frac{\log_e (1 + \mu |x|)}{\log_e (1 + \mu)} \tag{3.1}$$

where x is the normalized amplitude ($-1 \leq x \leq 1$) and y is the normalized coding range. In current applications in North America, μ is equal to 255.

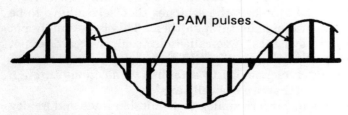

Figure 3.1   Sampling of an analog signal.

The A-law has a true linear segment for signals having an amplitude of less than $1/A$. The normalized expression of the A-law is

$$y = \frac{1 + \log_e (A\,|x|)}{1 + \log_e A} \qquad \text{for } \frac{1}{A} \leq |x| \leq 1 \tag{3.2}$$

$$= \frac{A\,|x|}{1 + \log_e A} \qquad \text{for } 0 \leq |x| \leq \frac{1}{A} \tag{3.3}$$

The standard value used today is $A = 87.6$.

A true representation in pulse code modulation of any analog signal will be obtained as long as the sampling frequency is twice the highest frequency in the analog signal according to the Nyquist theorem (1928). The voice signals transmitted in telephony are limited to a range of 300 to 3400 Hz, which is called a *nominal 4-kHz voice channel* for convenience. Twice this number gives a sampling rate of 8000 times per second.

Theoretically, any analog signal can be converted to a PCM signal as long as its sampling rate is twice that of the highest frequency in the analog signal. In compact discs (CDs), high-fidelity (hi-fi) signals with a frequency of up to 20,000 Hz are converted to digital signals at a sampling rate of 44,100 per second, one for each stereo channel. The samples are quantized and coded at 16 or 17 bits for each sample, resulting in a total bit rate of 1.5 Mbit/s. An analog high-definition television (HDTV) signal with a bandwidth of about 10 MHz would require a sampling rate of 20 million per second. Professor Mischa Schwartz (1990) has said that digitized HDTV will require a bit rate of 500 Mbit/s for each TV channel.

Some codings improve on the ratio of information to bits of code and thus can be considered a form of compression. One such code, adaptive differential pulse code modulation (ADPCM), is discussed in Sec. 3.4.3.

Quadrature amplitude modulation (QAM) is used for high-speed modems and to extend the bandwidth of copper-based subscriber loops. It belongs to a family of codes that increases the volume of information (e.g., number of bits) in relation to the bandwidth of the channel. In QAM, both the amplitude and the phase of a carrier signal are modulated to create a number of stages, typically 32 or 64, as shown by the constellations in Fig. 3.2. The amplitude of a signal is described by the distance from the origin and the phase by the angle with the horizontal axis.

Carrierless amplitude and phase (CAP) modulation is a variation of QAM that is used to increase the bit rate on copper-based subscriber loops. It is discussed in Sec. 4.3.

Much of the distortion on copper-based subscriber loops is frequen-

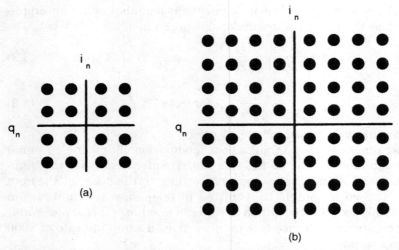

Figure 3.2 Quadrature amplitude modulation (QAM): (*a*) 16 QAM, (b) 64 QAM.

cy dependent in the sense that certain frequency bands carry more distortion than the rest. This opens up possibilities to increase the utility of copper-based loops. One such technique is based on *discrete multitone* (DMT) *channels*. The frequency band of a loop can be divided into 256 subchannels, each with a bandwidth of 4 kHz, as shown in Fig. 3.3. Through the use of digital signal processing (DSP), subchannels without or with insignificant distortion are selected and used for transmission of desired signals. Transmission parameters of subchannels with marginal distortion are adjusted by DSP to minimize errors and crosstalk. Low-frequency plain old telephone signals (POTS) can be sent under the mentioned 256 subchannels, as can certain ISDN signals. DMT is used for asymmetric digital subscriber line (ADSL), as described in Sec. 4.3.

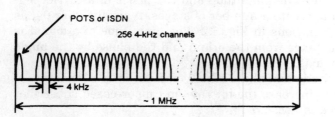

Figure 3.3 Discrete multitone channels.

3.4 Compression

Two different approaches can be attempted in communications: (1) to transmit information of the highest quality over existing channels or (2) to reduce the bandwidth required to transmit information of a given quality. Stated slightly differently, decisions also have to be taken regarding: (3) what information should be transmitted and (4) how should it be transmitted (Berger, 1971, p. 2). Compression of digital signals is a means of reducing the required transmission bit rate and plays a part in all four aspects. Compression is also used to reduce the volume of data stored in memories and on disks. Certain source data are highly redundant, particularly digitized images such as video and facsimile. Picture elements (pels) will be highly repetitive, particularly those representing white. If, for example, a digital signal contains a long string of *zeros,* it might be more economical to transmit a code indicating that a string of zeros follows and the length of that string. At the receiving end, the special code is converted back into the original string of zeros. In the same way, a string of *ones* can be represented by another code.

Many different, complex algorithms for compression and decompression of digital codes have been constructed. Some algorithms are listed in Table 3.3.

Other special algorithms are used for image and video compression. Among such algorithms are ISO-JBIG (Joint Bilevel Image Experts Group), ISO-JPEG (Joint Photographic Experts Group), and ISO-MPEG (Moving Picture Experts Group). JBIG applies to bilevel (black-and-white) images and such with limited bits per picture ele-

TABLE 3.3 Some Coding Algorithms

Input range	Transmission rate	Sampling rate	Bits/sample	Standard
Linear predictive coding algorithm	64,000 bit/s	2,400 bit/s		LPC-10
Code excited linear prediction (CELP)	7500 Hz	16,000 bit/s	16,000 Hz	
32 kbit/s adaptive differential pulse code modulation (ADPCM)	64,000 bit/s	32,000 bit/s		G.721
7-kHz audio coding within 64 kbit/s	50–7000 Hz	64,000 bit/s, 56,000 bit/s, 48,000 bit/s	16,000 Hz	G.722

ment, JPEG to still pictures, and MPEG to full motion picture images and video.

3.4.1 Run-length and pattern coding

Run-length coding is the formal name of techniques that substitute a short code for a long trail of symbols representing the same thing (such as a line of zeros or spaces).

In the CCITT Recommendation T.4 for group 3 facsimile machines, run-length coding is specified as a means of reducing the amount of information that has to be transmitted. An all-white line is represented by a code word that indicates the following number of white picture elements (pels). Several sets of code words are used, two to specify ranges of picture elements between 0 and 63, of which one is for runs of white picture elements and the other is for black ones, as shown in Tables 3.4 and 3.5, respectively. In a transmission, black-and-white code words alternate and two white or two black code words never follow each other. The two sets (for white and black) each individually meet the standards of Huffman coding discussed later. Two other sets extend the range of picture elements to 1728 (the maximum length of a scan line) by adding makeup codes of the type shown in Table 3.6. They add run-lengths of picture elements in groups of 64. By using the two in combination, any length of picture elements between 0 and 1728 can be encoded using two code words.

TABLE 3.4 Terminating Codes for Group 3 and 4 Facsimiles—White Runs

Run length	Code word	Run length	Code word	Run length	Code word	Run length	Code word
0	00110101	16	101010	32	00011011	48	00001011
1	000111	17	101011	33	00010010	49	01010010
2	0111	18	0100111	34	00010011	50	01010011
3	1000	19	0001100	35	00010100	51	01010100
4	1011	20	0001000	36	00010101	52	01010101
5	1100	21	0010111	37	00010110	53	00100100
6	1110	22	0000011	38	00010111	54	00100101
7	1111	23	0000100	39	00101000	55	01011000
8	10011	24	0101000	40	00101001	56	01011001
9	10100	25	0101011	41	00101010	57	01011010
10	00111	26	0010011	42	00101011	58	01011011
11	01000	27	0100100	43	00101100	59	01001010
12	001000	28	0011000	44	00101101	60	01001011
13	000011	29	00000010	45	00000100	61	00110010
14	110100	30	00000011	46	00000101	62	00110011
15	110101	31	00011010	47	00001010	63	00110100

TABLE 3.5   Terminating Codes for Group 3 and 4 Facsimiles—Black Runs

Run length	Code word	Run length	Code word	Run length	Code word	Run length	Code word
0	0000110111	16	0000010111	32	000001101010	48	000001100100
1	010	17	0000011000	33	000001101011	49	000001100101
2	11	18	0000001000	34	000011010010	50	000001010010
3	10	19	00001100111	35	000011010011	51	000001010011
4	011	20	00001101000	36	000011010100	52	000000100100
5	0011	21	00001101100	37	000011010101	53	000000110111
6	0010	22	00000110111	38	000011010110	54	000000111000
7	00011	23	00000101000	39	000011010111	55	000000100111
8	000101	24	00000010111	40	000001101100	56	000000101000
9	000100	25	00000011000	41	000001101101	57	000001011000
10	0000100	26	000011001010	42	000011011010	58	000001011001
11	0000101	27	000011001011	43	000011011011	59	000000101011
12	0000111	28	000011001100	44	000001010100	60	000000101100
13	00000100	29	000011001101	45	000001010101	61	000001011010
14	00000111	30	000001101000	46	000001010110	62	000001100110
15	000011000	31	000001101001	47	000001010111	63	000001100111

The code words range in size from 2 to 13 bits. Code words are also available that increase the length of a scan line to 2560 pels, as shown in Table 3.7. They are intended for paper sizes wider than 215 mm or $8^{15}/_{32}$ in.

The facsimile coding specified in CCITT Recommendation T.4 is designed for a maximum of 1728 pels along a 215-mm ($8^{15}/_{32}$-in) line corresponding to 8 pels/mm or 204 pels/in. This is low compared with present resolutions. Enhanced facsimile terminals can already

TABLE 3.6   Makeup Codes for Group 3 and 4 Facsimiles

White runs				Black runs			
Run length	Code word	Run length	Code word	Run length	Code word	Run length	Code word
64	11011	960	011010100	64	0000001111	960	0000001110011
128	10010	1024	011010101	128	000011001000	1024	0000001110100
192	010111	1088	011010110	192	000011001001	1088	0000001110101
256	0110111	1152	011010111	256	000001011011	1152	0000001110110
320	00110110	1216	011011000	320	000000110011	1216	0000001110111
384	00110111	1280	011011001	384	000000110100	1280	0000001010010
448	01100100	1344	011011010	448	000000110101	1344	0000001010011
512	01100101	1408	011011011	512	0000001101100	1408	0000001010100
576	01101000	1472	010011000	576	0000001101101	1472	0000001010101
640	01100111	1536	010011001	640	0000001001010	1536	0000001011010
704	011001100	1600	010011010	704	0000001001011	1600	0000001011011
768	011001101	1664	011000	768	0000001001100	1664	0000001100100
832	011010010	1728	010011011	832	0000001001101	1728	0000001100101
896	011010011	EOL	000000000001	896	0000001110010	EOL	000000000001

TABLE 3.7   Additional Makeup Codes—
White and Black

Run length	Makeup code words
1792	00000001000
1856	00000001100
1920	00000001101
1984	000000010010
2048	000000010011
2112	000000010100
2176	000000010101
2240	000000010110
2304	000000010111
2368	000000011100
2432	000000011101
2496	000000011110
2560	000000011111

increase the resolution by using the additional makeup codes on paper that is 215 mm ($8^{15}\!/_{32}$ in) wide.

Pattern coding is a method of reducing the length of text. A single code word substitutes for patterns of 2 or more bytes or other pieces of information. The meaning of the code words is kept in a code table, copies of which are kept at each end of the transmission link. As long as the selected code word is shorter than the combined code words for the individual characters, a reduction in total message length is obtained. This reduction can be maintained as long as the transmission of the code tables between encoder and decoder does not consume the entire gain. The number of bits in a code word is made dependent on the frequency of occurrence of a particular pattern of bytes in such a way that frequent patterns of bytes are allocated a short code word and infrequent ones are represented with increasingly longer code words. The type of coding used is a modified Huffman coding, a form of variable-length coding. These concepts are discussed in more detail below. Variable-length coding is also called *entropy coding* in the literature.

3.4.2   Variable-length character coding

3.4.2.1   Morse code.   Probably the first example of variable-length character coding is the Morse code. It was invented by Samuel Finley Breese Morse in 1844. It uses a short code for frequently used characters and longer codes for characters that are used infrequently. The most frequently used letter in English literature is the *E,* with an occurrence of 13 percent. In the Morse code it is represented by a dot. The least frequent letters are *Q* and *Z,* which occur ¼ percent of the

TABLE 3.8   Morse Code

A	·—	B	—···	C	—·—·	D	—··
E	·	F	··—·	G	——·	H	····
I	··	J	·————	K	—·—	L	·—··
M	——	N	—·	O	———	P	·——·
Q	——·—	R	·—·	S	···	T	—
U	··—	V	···—	W	·——	X	—··—
Y	—·——	Z	——··	1	·————	2	··———
3	···——	4	····—	5	·····	6	—····
7	——···	8	———··	9	————·	0	—————
Period			·—·—·—	Comma			——··——

time. They are represented by the codes — — · — and — — · · ,
respectively. Table 3.8 shows the Morse code.

3.4.2.2 Huffman coding. *Huffman coding* is a form of a variable-length word code. It is used mainly to compress text. It was invented by David Huffman in 1952.

When a variable-length character code is used to create a string of bits, it is important that no character code appears as a prefix in another character code. If this rule is adhered to, there will be no ambiguity in decoding the string. For example, if the code for the letter *a* is *01*, no other letter code may begin with the sequence *01*. This will be true if the character codes belong to a *binary tree*. Figure 3.4 shows binary trees as used for Huffman coding.

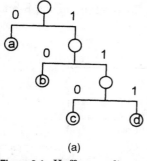

(a)

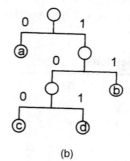

(b)

Figure 3.4 Huffman coding tree.

TABLE 3.9   Coding Tree

Left tree, (a)		Right tree, (b)	
a	0	a	0
b	10	b	110
c	110	c	100
d	111	d	101

At each fork the left branch is represented by the binary digit *0* and the right branch by the digit *1*. When a branch ends with a leaf, a character is represented. The coding in Table 3.9 is represented by the two different trees, (a) and (b) to the left and right, respectively.

Within each of the two tree examples, no two codes have the same prefix, and the string can be decoded without ambiguity. In the preceding sentence, characters appear with the frequency shown in Table 3.10. Instead of the frequency of appearance, the probability of appearance can be used.

Figure 3.5 shows an example of a Huffman coding tree for the character frequencies shown in Table 3.10. It starts with the least frequent characters, l, v, W, y, comma, and period (which each appear once) in the lower left corner of the tree. The numbers to the right of the node above the leaves indicate the sum of the frequencies of the character leaves below. The number of levels is minimized by always combining branches with the lowest frequencies.

Figure 3.6 resembles Figure 3.5, but it has been changed to show the coding steps taken to reach each leaf, i.e., the coding to each leaf. Table 3.11 lists the same information in table form.

As seen in Table 3.11, the most frequent symbol, the space, is coded with 2 bits, while the least frequent ones are coded with 7 bits each. By multiplying the number of bits by the number of times the character appears in the message, we obtain the total length of the message in bits. In this case, it is 507 bits, or 4.26 bits per character. This is 61 percent of the bits in a 7-bit ASCII code and 53 percent of an 8-bit

TABLE 3.10   Frequency Table

Space	21	d	5	b	2	v	1
e	15	n	5	f	2	W	1
t	10	c	4	g	2	y	1
a	7	s	4	p	2	Comma	1
h	7	m	3	u	2	Period	1
i	7	r	3	x	2		
o	7	w	3	l	1		

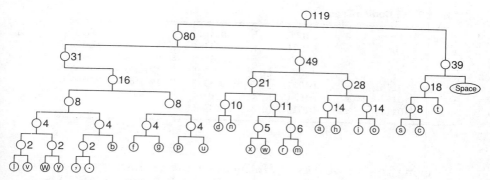

Figure 3.5 Example of Huffman coding.

ASCII code. These correspond to compression values of 39 and 47 percent, respectively. Compressions with Huffman coding on the order of 30 to 55 percent are reported in the literature.

The coding scheme has to be known to both the coder and the decoder. This is no problem if the Huffman coding is based on a standard frequency table of the characters. Such tables are available for all languages, as well as for registers of languages (scientific, colloquial, slang, etc.). In these cases, a fixed, standard Huffman code can be obtained and used in the coders and decoders. However, in most messages, the frequency of the characters used differs from message to message. In order to maximize the compression, different compression tables based on the character frequency tables of each message should be used. This means that new coding information has to be transmitted from the coders to the decoders for each message. Obviously, the transmission of such coding information increases the length of the material transmitted and can consume the gain obtained by the compression.

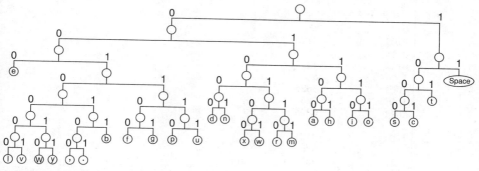

Figure 3.6 Detailed coding of preceding example.

TABLE 3.11   Coding Table

Space	11	d	01000	b	001011	v	0010001
e	000	n	01001	f	001100	W	0010010
t	101	c	1001	g	001101	y	0010011
a	01100	s	1000	p	001110	Comma	0010100
h	01101	m	010111	u	001111	Period	0010101
i	01110	r	010110	x	010100		
o	01111	w	010101	l	0010000		

Among the drawbacks of Huffman coding is that loss of one bit results in loss of the entire message. If a bit is lost in an ASCII code, just one character is lost.

3.4.2.3   Arithmetic coding.   *Arithmetic coding* is another form of variable-length coding. The code can be viewed as fractional binary values between 0 and 1 on a number line. A binary fractional number is commonly represented similarly to a decimal number (or fraction). The first position to the right of a dot corresponds to $\frac{1}{2}$ in the base-10 system. Similarly, the second position to the right of the dot represents $\frac{1}{4}$, the third $\frac{1}{8}$, and so on. Thus the fractional binary number .101 is equal to $\frac{1}{2} + \frac{1}{8}$, or .625 in the base-10 system. Similarly, .0101 equals $\frac{1}{4} + \frac{1}{16}$, or .3125. Table 3.12 shows some fractional binary numbers.

TABLE 3.12   Fractional Binary Numbers

Fractional binary number	Decimal number
0	0
.0001	.0625
.001	.125
.0011	.1875
.01	.250
.0101	.3125
.011	.375
.0111	.4375
.1	.5
.1001	.5625
.101	.625
.1011	.6875
.11	.75
.1101	.8125
.111	.875
.1111	.9375
1	1

TABLE 3.13   Example 1 of Arithmetic Coding

| Symbol | Frequency | Probability | | Cumulative | |
		Decimal	Binary	Low	High
α	32	0.5	0.1	0	0.1111111
β	16	0.25	0.01	0.1	0.01111111
χ	8	0.125	0.001	0.11	0.11011111
δ	8	0.125	0.001	0.111	0.11111111
TOTAL	64	1			

To simplify the description of arithmetic coding, symbol frequencies that are a power of 2 have been used, and the total length of the sample is also a power of 2. Table 3.13 shows the frequencies of some symbols. First, they are expressed in decimal form and then in fractional binary numbers. The cumulative values of the binary probabilities starting with zero are shown in the fifth column, marked "Low cumulative probability of binary numbers."

Figure 3.7 shows a graphic representation of the same coding. A line between the binary numbers zero (0) and one (1) is subdivided according to Table 3.13. By convention, the symbol for α is given the interval between 0.00 and close to but not including 0.01. Similarly, the symbol for the character β has an interval that includes 0.01 and approaches 0.10.

By convention, any number in an interval can be chosen. Thus, in the case of the α symbol, any of the fractional binary numbers between 0 and 0.01 are acceptable. Since we are dealing with compression, the shortest number, i.e., 0, should be selected. Generally speaking, this means that we select the lowest number. We can change this convention and select the highest number instead, i.e., 0.10, 0.110, 0.111, and 1.00.

Table 3.14 was intended to include all the 119 symbols in the original example, i.e., *"Within each of the two tree examples, no two codes have the same prefix, and the string can be decoded without ambiguity."* However, the total number of symbols has been increased to 128 and the frequencies of some frequencies modified in order to meet the requirements of restricting the numbers in arithmetic coding to pow-

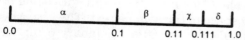

Figure 3.7 Example of arithmetic coding.

TABLE 3.14   Example 2 of Arithmetic Coding

		Probability			
				Cumulative	
Symbol	Frequency	Decimal	Binary	Low	High
Space	32	0.25	0.01	0.00	0.00111111
e	16	0.125	0.001	0.010	0.01011111
t	16	0.125	0.001	0.011	0.01111111
a	8	0.0625	0.0001	0.1000	0.10001111
h	8	0.0625	0.0001	0.1001	0.10011111
i	4	0.03125	0.00001	0.10100	0.10100111
o	4	0.03125	0.00001	0.10101	0.10101111
d	4	0.03125	0.00001	0.10110	0.10110111
n	4	0.03125	0.00001	0.10111	0.10111111
c	4	0.03125	0.00001	0.11000	0.11000111
s	4	0.03125	0.00001	0.11001	0.11001111
m	2	0.015625	0.000001	0.110100	0.11010011
r	2	0.015625	0.000001	0.110101	0.11010111
w	2	0.015625	0.000001	0.110110	0.11011011
b	2	0.015625	0.000001	0.110111	0.11011111
f	2	0.015625	0.000001	0.111000	0.11100011
g	2	0.015625	0.000001	0.111001	0.11100111
p	2	0.015625	0.000001	0.111010	0.11101011
u	2	0.015625	0.000001	0.111011	0.11101111
x	2	0.015625	0.000001	0.111100	0.11110011
l	1	0.0078125	0.0000001	0.1111010	0.11110101
v	1	0.0078125	0.0000001	0.1111011	0.11110111
W	1	0.0078125	0.0000001	0.1111100	0.11111001
y	1	0.0078125	0.0000001	0.1111101	0.11111011
Comma	1	0.0078125	0.0000001	0.1111110	0.11111101
Period	1	0.0078125	0.0000001	0.1111111	1.00000000
				1.0000000	
TOTAL	128	1			

ers of 2. As the number of symbols to be coded increases, the interval for each symbol becomes smaller. This means that the code symbols become longer.

Arithmetic coding is based on the probability p of the occurrence of a symbol and the accumulated probabilities P. The first coding point, C_0, is usually set to zero. The first interval between coding points, A_0, is set to 1.; i.e., it goes from coding point 0. to point 1.0. The next interval, A_1, is calculated as the previous interval times the probability of the new symbol, i.e., $A_0 \times p_1$. The new coding point, C_1, is calculated as C_0 plus the *augend*, $A_0 \times P_1$. Table 3.15 shows the arithmetic coding of the first word and the following space of the sample text. The accuracy of the p's and P's has been limited to 7 bits after the binary point. In

TABLE 3.15   Arithmetic Coding of the Example "Within each..."

	$C_0 =$	0 $A_0 = 1.$
W	$n = 1$	$p_1 = .0000001$ $P_1 = .1111100$
	$A_1 = A_0 \times p_1 =$	$1. \times .0000001 = .0000001$
	$A_0 \times P_1 =$	$1. \times .1111100 = .1111100$
	$C_1 = C_0 + (A_0 \times P_1) =$	$0 + .1111100 = .1111100$
i	$n = 2$	$p_2 = .00001$ $P_2 = .10100$
	$A_2 = A_1 \times p_2 =$	$.0000001 \times .00001 = .000000000001$
	$A_1 \times P_2 =$	$.0000001 \times .10100 = .000000010100$
	$C_2 = C_1 + (A_1 \times P_2) =$	$.1111100 + .000000010100 = .111110010100$
t	$n = 3$	$p_3 = .001$ $P_3 = .011$
	$A_3 = A_2 \times p_3 =$	$.000000000001. \times 001 = .000000000000001$
	$A_2 \times P_3 =$	$.000000000001 \times .011 = .000000000000011$
	$C_3 = C_2 + (A_2 \times P_3) =$	$.111110010100 + .000000000000011 = .111110010100011$
h	$n = 4$	$p_4 = .0001$ $P_4 = .1001$
	$A_4 = A_3 \times p_4 =$	$.000000000000001 \times .0001 = .0000000000000000001$
	$A_3 \times P_4 =$	$.000000000000001 \times .1001 = .0000000000000001001$
	$C_4 = C_3 + (A_3 \times P_4) =$	$.111110010100011 + 0000000000000001001 =$
		$.1111100101000111001$
i	$n = 5$	$p_5 = .00001$ $P_5 = .10100$
	$A_5 = A_4 \times p_5 =$	$.0000000000000000001 \times .00001 = .000000000000000000000001$
	$A_4 \times P_5 =$	$.0000000000000000001 \times .10100 = .000000000000000000010100$
	$C_5 = C_4 + (A_4 \times P_5) =$	$.1111100101000111001 + .000000000000000000010100 =$
		$.111110010100011100110100$
n	$n = 6$	$p_6 = .00001$ $P_6 = .10111$
	$A_6 = A_5 \times p_6 =$	$.000000000000000000000001 \times .00001 =$
		$.00000000000000000000000000001$
	$A_5 \times P_6 =$	$.000000000000000000000001 \times .10111 =$
		$.00000000000000000000000010111$
	$C_6 = C_5 + (A_5 \times P_6) =$	$.111110010100011100110100 +$
		$.00000000000000000000000010111 =$
		$.11111001010001110011010010111$
Space	$n = 7$	$p_7 = .01$ $P_7 = .00000001$
	$A_7 = A_6 \times p_7 =$	$.00000000000000000000000000001 \times .01 =$
		$.0000000000000000000000000000001$
	$A_6 \times P_7 =$	$.00000000000000000000000000001 \times .00000001 =$
		$.00000000000000000000000000000000000001$
	$C_7 = C_6 + (A_6 \times p_7) =$	$.11111001010001110011010010111 +$
		$.00000000000000000000000000000000000001 =$
		$.11111001010001110011010010111100000001$

the case of space, the cumulative probability P equals zero. Using this value would make the augend equal to zero, and there would be no difference in the coding of the space and the previous symbol. Instead of zero, space has been given the lowest cumulative probability; i.e., P_{space} = .0000001. Any value between the lowest and highest cumulative probabilities can be selected for the coding of a symbol.

Another approximation is recommended in the Joint Photographic Experts Group (JPEG) standard. It uses the probability estimate Qe for the less probable symbol (LPS) and $1 - Qe$ for the more probable symbol (MPS) as per Pennebaker and Mitchell (1992). Here, $1 - Qe = P$, and $Qe = 1 - P$. The exact formula then becomes

$$C_n = C_{n-1} + (A_{n-1} \times P_n) = C_{n-1} + A_{n-1} \times (1 - Qe_n) = C_{n-1} + A_{n-1} - Qe_n$$

which is approximated by

$$C_n = C_{n-1} + A_{n-1}$$

This avoids multiplication with zero when $P_n = 0$.

Coding of the 119 symbols in the example shown in Table 3.10 according to the arithmetic method requires 487 bits. As mentioned earlier, with Huffman coding the total number of bits required is 507. With a 7-bit ASCII code, 833 bits are required. Thus, in this particular example, arithmetic coding offers a slight advantage compared with Huffman coding.

Decoding of arithmetic coding is essentially performed through calculations that are the reverse of the ones used in coding. Let us look at the coding string after the first 7 symbols have been coded, that is,

$$C_7 = .111110010100011100110100101110000001 \qquad (3.4)$$

This string lies between the values

$$C = .1111100 \quad \text{and} \quad C = .11111001 \qquad (3.5)$$

According to Table 3.14, this interval represents the symbol W, and we thus know that the first symbol in the sentence is W. From Table 3.14 we also know the values of p_W and P_W, that is, .0000001 and .1111100, respectively.

Now we subtract P_W from the value of the coding string, that is,

$$\begin{aligned} .111110010100011100110100101110000001 - \\ .1111100 = \qquad (3.6) \\ .000000010100011100110100101110000001 \end{aligned}$$

and divide this number by the value for p_W, that is, .0000001. The result is

$$.101000111001101001011100000001 \qquad (3.7)$$

which falls in the interval between .10100 and .10100111. This means that the string represents the symbol i.

Table 3.16 shows the complete decoding of the first part of the sample sentence.

The additive nature of arithmetic coding sometimes leads to a *carry-over problem*. Langdon (1984, pp. 139 and 144) discusses this and how to control it.

3.4.3 Adaptive differential pulse code modulation (ADPCM)

Variations of pulse code modulation (PCM) include differential pulse code modulation (DPCM) and delta modulation (DM). In these types of modulation, the differences between previous and current samples are encoded and transmitted rather than the full value of the current sample. This way less data have to be transmitted. In DPCM, the value of the difference between the previous and current sample is measured, encoded, and transmitted. Some forms of DPCM look at several past samples to calculate the difference between those and the current sample. DM is a special form of DPCM in which a single bit is used to indicate the differences between samples. If the current sample is larger than the preceding sample, a *one* is transmitted; if it is smaller, a *zero* is transmitted. In conventional DPCM, the weighting coefficients assigned to each past sample are fixed. In *adaptive coding,* they are adjustable. The values of the weighting coefficients are adjusted depending on the size of the fluctuations in the difference signals.

The CCITT has adopted Recommendation G.721, which specifies the compression of conventional 64-kbit/s PCM signals to 32 kbit/s using adaptive differential pulse code modulation (ADPCM). In this approach, the scaling factor of the quantizer is adapted in a fast and a slow mode. The fast mode is for signals with large fluctuations, such as speech, and the slow mode is for signals with small fluctuations, such as voiceband data and tones.

Figure 3.8 shows a block diagram of an ADPCM encoder. PCM signals at the bit rate of 64 kbit/s in either the A-law or μ-law format are presented to the encoder. The algorithm in the first block converts the input signal to a uniform PCM signal by removing the A-law or μ-law compression. The next algorithm compares the uniform signal with an estimated signal and calculates the difference. An *adaptive quan-*

TABLE 3.16   Decoding of the Sample Text

Original string, C_7 =	.111110010100011100110100101 1100000001		
Interval	.1111100	.11111001	W
Subtract P_W	.1111100		
Intermediate result	.00000001010001110011010010 11100000001		
Divide by p_W	.0000001		
Intermediate result	.101000111001101001011100000001		
Interval	.101000	.10100111	i
Subtract P_i	.10100		
Intermediate result	.00000011100110100101110000 0001		
Divide by p_i	.00001		
Intermediate result	.011100110100101 1100000001		
Interval	.011	.01111111	t
Subtract P_t	.011		
Intermediate result	.000100110100101 1100000001		
Divide by p_t	.001		
Intermediate result	.1001101001011100000001		
Interval	.1001	.10011111	h
Subtract P_h	.1001		
Intermediate result	.0000101001011100000001		
Divide by p_h	.0001		
Intermediate result	.101001011100000001		
Interval	.10100	.10100111	i
Subtract P_i	.10100		
Intermediate result	.000001011100000001		
Divide by p_i	.00001		
Intermediate result	.1011100000001		
Interval	.10111	.10111111	n
Subtract P_n	.10111		
Intermediate result	.0000000000001		
Divide by p_n	.00001		
Intermediate result	.0000001		
Interval	.00	.00111111	Space

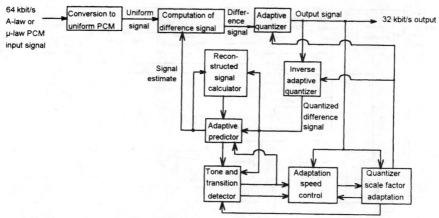

Figure 3.8 Adaptive differential pulse code modulation encoder.

tizer, of the 15-level nonuniform type, converts the difference signal to a base 2 logarithmic representation and scales it by the input from a quantizer scale factor adapter. Then it quantizes the difference signal using a 4-bit representation (3 for magnitude and 1 for the sign) and produces the 32-kbit/s *output signal.* This signal is also fed to the inverse adaptive quantizer, the quantizer scale factor adaptation, and the adaptation speed control.

The *inverse adaptive quantizer* produces a quantized version of the difference signal by scaling and applying the output from the quantizer scale factor adapter and special factors from a look-up table. It also transforms the result from the logarithmic format. The *quantizer scale factor adaptation algorithms* compute the scaling factors for the adaptive quantizer and the inverse adaptive quantizer. These computations are based on the 4-bit output from the adaptive quantizer and information from the adaptation speed control. This unit operates in two different modes, fast and slow. The fast adaptation mode is used for signals with large fluctuations, such as speech, and the slow adaptation mode is used for signals with small fluctuations, such as voiceband data and tones. Parameters from both modes are combined to control the speed of adaptation. The function of the tone and transition detector is to improve the performance for signals from modems of the frequency-shift (FSK) type that operate in the symbol mode. It forces the adaptive quantizer into the fast mode of adaptation.

The adaptive predictor computes the estimate of the input signal with the assistance of the reconstructed signal calculator using elaborate algorithms from CCITT Recommendation G.721.

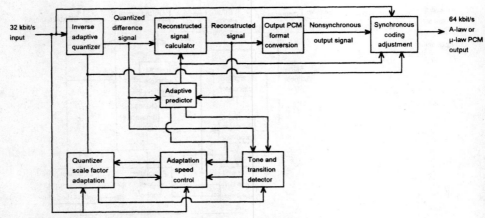

Figure 3.9 Adaptive differential pulse code modulation decoder.

Figure 3.9 shows a block diagram of the corresponding ADPCM decoder. All the blocks except the one performing the synchronous coding adjustment perform the same functions as in the encoder. The block with the synchronous coding adjustment prevents distortions from accumulating on synchronous links with a series of conversions from PCM to ADPCM and back.

3.4.4 Modified READ (relative element address designate) coding

When transmitting lines of picture element information, as in facsimile transmission, it is often economical to transmit changes from a previous line. The reason for this is that successive scan lines tend to be similar. One algorithm for this is called *modified READ (relative element address designate) coding*. Run-length coding that uses only one line at a time is called *one-dimensional coding*. In comparison,

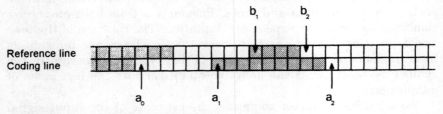

Figure 3.10 Definitions of changing picture elements.

modified READ coding, which uses a previous line, is called *two-dimensional coding.*

In modified READ (MR) coding, the first scan line is coded in modified Huffman (MH) code of the type shown in Tables 3.4 through 3.7. The picture elements of this line are retained in memory, and the line becomes the *reference line,* as shown in Fig. 3.10. The new line being coded is called the *coding line.* Black picture elements (pels) are shown in a striped pattern in Fig. 3.10. White picture elements are shown in white. A changing picture element is one that changes from white to black or vice versa in a scan. This happens with the picture elements marked b_1 and b_2 in the reference line and with those marked a_0, a_1, and a_2 in the coding line. The definitions of the changing elements are as follows according to CCITT Recommendations T.4 and T.6:

a_0 The reference or starting changing element on the coding line

a_1 The next changing element to the right of a_0 on the coding line

a_2 The next changing element to the right of a_1 on the coding line

b_1 The first changing element on the reference line to the right of a_0 that is of opposite "color" (white versus black or black versus white) to a_0

b_2 The next changing element to the right of b_1 on the reference line

When there is a change in element "color" in the reference line but not in the coding line (i.e., when b_2 appears to the left of a_1, as shown in Fig. 3.11), it is passed by. This is called the *pass-mode code.* If a_1 occurs just under b_2, this is *not* a pass-mode state.

A *vertical mode* is identified when the position of a_1 in relation to b_1 is equal to or less than 3 pels. Figure 3.12 shows an example where a_1 is 3 pels to the left of b_1. Figure 3.13 shows another example of a vertical mode. In this case, a_1 is 1 pel to the left of b_1. A third example is shown in Fig. 3.14, where a_1 is 2 pels to the right of b_1. For each of the seven possibilities of the vertical mode, a unique code word is used, as shown in Table 3.17.

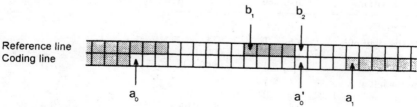

Figure 3.11 Pass-mode coding.

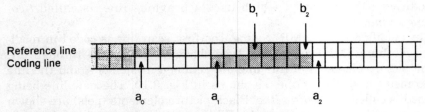

Figure 3.12 Vertical mode with a_1 3 pels to the left of b_1.

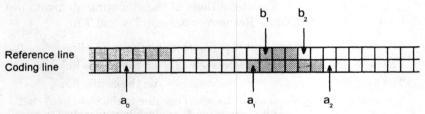

Figure 3.13 Vertical mode with a_1 1 pel to the left of b_1.

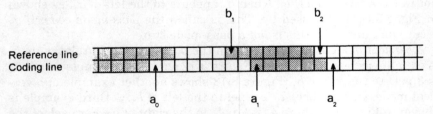

Figure 3.14 Vertical mode with a_1 2 pels to the right of b_1.

TABLE 3.17   Coding Table for the Vertical Mode

Position	Distance	Code word
a_1 just under b_1	$a_1b_1 = 0$	1
a_1 to the right of b_1	$a_1b_1 = 1$	011
	$a_1b_1 = 2$	000011
	$a_1b_1 = 3$	0000011
a_1 to the left of b_1	$a_1b_1 = 1$	010
	$a_1b_1 = 2$	000010
	$a_1b_1 = 3$	0000010

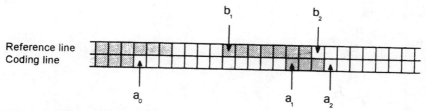

Figure 3.15 Horizontal mode with a_1 5 pels to the right of b_1.

The situation where the distance between a_1 and b_1 is more than 3 pels is called *horizontal mode*. Figure 3.15 shows an example where a_1 is positioned 5 pels to the right of b_1. Another example of a horizontal mode is shown in Fig. 3.16. In this case, a_1 is 4 pels to the left of b_1. In the horizontal mode, the coding line is coded according to the modified Huffman code. In order to limit the area that is distorted by a coding error, at least every second or fourth line must be coded according to the one-dimensional mode.

3.4.5 Discrete cosine transform

By transforming a sequence of symbols to another format, one can sometimes compress the sequence. One such transform is the *discrete cosine transform* (DCT). It is similar to the Fourier transform but only uses the cosine function. Another difference is that DCT only deals with discrete values, not with continuous ones. This transform function was first described by Ahmed et al. in 1974. The DCT of a sequence

$$f(x), \qquad m = 0, 1, \ldots, (M - 1) \tag{3.8}$$

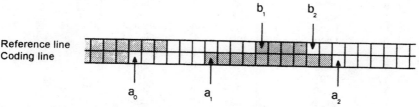

Figure 3.16 Horizontal mode with a_1 4 pels to the left of b_1.

is defined as

$$g_x(0) = \frac{\sqrt{2}}{M} \sum_{m=0}^{M-1} f(x) \tag{3.9}$$

$$g_x(k) = \frac{2}{M} \sum_{m=0}^{M-1} f(m) \cos \frac{(2m+1)k\pi}{2M}, \qquad k = 1, 2, \ldots, (M-1) \tag{3.10}$$

The inverse discrete cosine transform (IDCT) is defined as

$$f(m) = \frac{1}{\sqrt{2}} g_x(0) + \sum_{k=1}^{M-1} g_x(k) \cos \frac{(2m+1)k\pi}{2M},$$
$$m = 0, 1, \ldots, (M-1) \tag{3.11}$$

Let us use the sequence C_7 in Table 3.15 as an example. It contains 37 bits; that is, $M = 37$. These bits are listed as 1s and 0s in the $f_1(x)$ column in Table 3.18. The corresponding $g(x)$ values are calculated using Eqs. (3.9) and (3.10). Applying Eq. (3.11), the $g(x)$ values are converted back into $f(x)$ values and shown in column $f_2(x)$.

The stated purpose of using the DCT is to compress the sequence; i.e., the number of $g(x)$ values should be less than M. As shown in Table 3.18, 20 values of $g(x)$ are close to 0 and in the $-.1 < g(x) < .1$ range. Table 3.19 shows the resulting $f_2(x)$ values, when the $g(x)$ values in this range are set equal to zero.

TABLE 3.18   DCT Example 1

m	$f_1(x)$	$g(x)$	$f_2(x)$	m	$f_1(x)$	$g(x)$	$f_2(x)$
0	1	.6880	1	19	1	.0216	1
1	1	.1866	1	20	0	.0955	0
2	1	−.0080	1	21	1	.0786	1
3	1	.1803	1	22	0	−.0237	0
4	1	−.0997	1	23	0	−.0554	0
5	0	.1114	0	24	1	.1414	1
6	0	.1132	0	25	0	−.1555	0
7	1	−.1603	1	26	1	−.1880	1
8	0	.1539	0	27	1	.0022	1
9	1	−.1065	1	28	1	.0731	1
10	0	.0630	0	29	0	−.1026	0
11	0	.0809	0	30	0	.0869	0
12	0	−.1364	0	31	0	.1694	0
13	1	−.2356	1	32	0	−.0055	0
14	1	−.0397	1	33	0	.0275	0
15	1	.0286	1	34	0	.0412	0
16	0	.2157	0	35	0	−.1859	0
17	0	−.0964	0	36	1	−.0745	1
18	1	.0046	1				

TABLE 3.19   DCT Example 2

m	$f_1(x)$	$g(x)$	$f_2(x)$	m	$f_1(x)$	$g(x)$	$f_2(x)$
0	1	.6880	.7176	19	1	.0000	.7320
1	1	.1866	1.0801	20	0	.0000	−.0162
2	1	.0000	.8400	21	1	.0000	.7748
3	1	.1803	1.1512	22	0	.0000	.1257
4	1	.0000	1.0555	23	0	.0000	.0035
5	0	.1114	.1235	24	1	.1414	.9081
6	0	.1132	.0096	25	0	−.1555	.2007
7	1	−.1603	.6330	26	1	−.1880	1.2130
8	0	.1539	.2272	27	1	.0000	1.0278
9	1	−.1065	1.3918	28	1	.0000	.9904
10	0	.0000	0.0289	29	0	−.1026	−.0996
11	0	.0000	−.0181	30	0	.0000	.3349
12	0	−.1364	.2716	31	0	.1694	.0466
13	1	−.2356	.9459	32	0	.0000	−.0244
14	1	.0000	.7216	33	0	.0000	−.0422
15	1	.0000	.6986	34	0	.0000	−.1973
16	0	.2157	.3868	35	0	.1859	.1007
17	0	.0000	−.1202	36	1	.0000	.8410
18	1	.0000	.9359				

The largest $f_2(x)$ value that should be 0 in Table 3.19 is 0.39, and the smallest $f_2(x)$ value that should be 1 is 0.63. Thus the spread is large enough to distinguish between 0s and 1s without any problem with proper filtering. Thus only 17 of 37 $g(k)$ values have to be transmitted, a compression ratio of 0.54. If we set the $g(k)$ values in the $-.15 < g(k) < .15$ range to 0, only 11 of 37 values have to be transmitted, and the compression is increased to 0.70. However, in this case, the $f_2(x)$ values for 0s and 1s overlap, and the procedure does not work.

The discrete cosine transformation (DCT) can be combined with any other compression algorithm, as in the preceding example. Several such applications of DCT are discussed below.

3.4.6   Still picture coding

So far we have discussed the compression of text and digital facsimile pictures. The latter are bilevel or binary in the sense that a picture consists of black dots on a white background. There is no gray scale. The inclusion of a range of gray increases the required number of bits per picture element (pel). The addition of different colors further increases the required bits. The efficiency of compression can be increased by taking the grayness and/or color of neighboring picture elements into account. A group called the *Joint Photographic Experts Group* (JPEG) has been working on a standard for many years. This standard was approved by the International Organization for Standardization (ISO) in November of 1992 under the title "Digital Compression and Coding of Continuous-Tone Still Images." It is published in two parts: Part 1, "Requirements and Guidelines," as document IS-10918-1, and Part 2, "Compliance Testing," as document IS-10918-2. These standards are also published by CCITT as Recommendations T.81 and T.83, respectively.

The JPEG standard includes the following features:

1. Different compression-to-quality ratios can be set.
2. It can be applied to all continuous-tone digital source images.
3. It is easy to use in any hardware and/or software environment.
4. It contains the following modes of operation:
   a. Sequential encoding
   b. Progressive encoding
   c. Lossless encoding
   d. Hierarchical encoding

In most cases, a higher compression rate means a lower quality of the received picture. In other words, there is a compromise between compression and quality. The JPEG standard offers the possibility of selecting a desired rate of quality versus compression. This compression standard can be used for any continuous-tone (gray-scale) digital images from any source, as well as for color images. The JPEG standard is transparent to most types of hardware and software environments.

The JPEG compression standard offers four different modes of operation. In the *sequential encoding mode,* each picture element (pel) is encoded sequentially in a single scan from left to right and from top to bottom. In a mode referred to as *progressive encoding,* the imaging is performed in multiple scan passes, and the received image is built up over a relatively long time from a coarse image to one that becomes clearer for each scan pass. Thus the viewer sees the image being built up. With *lossless encoding,* the exact recovery of every feature of the image is guaranteed. The cost is a relatively low compression ratio. *Hierarchical encoding* refers to a mode in which the original image is encoded in several versions with different resolutions. At the receiving end, the required resolution can be selected, and the decompressions of versions with higher resolutions are left out. This saves the decoder from having to decompress versions with an unnecessarily high resolution. Each of the four modes of operation requires distinct versions of the coder-decoder (codec) pair. It is not expected that all applications will use all four modes of operation. Further, in some applications only one of the codec pairs will be required. Of the standardized modes of operation and image coding techniques, *baseline discrete cosine transform* (DCT) *sequential coding* is the simplest.

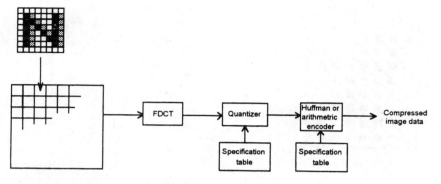

FDCT = forward discrete cosine transform

Figure 3.17 DCT encoder.

In this technique, the source image is sampled in blocks consisting of 8 × 8 pels each. Figure 3.17 shows a baseline DCT encoder. Each 8 × 8-pel block represents eight values along an x axis and eight along a y axis, and the block can be represented as a 64-point discrete signal. This signal can be considered as a spatial frequency domain input signal to a forward discrete cosine transform (FDCT). The output from the FDCT is 64 "DCT coefficients," consisting of 1 DC coefficient and 63 AC coefficients. The mathematical expression of the FDCT is

$$F(u,v) = \frac{1}{4} C(u)C(v) \left[\sum_{x=0}^{7} \sum_{y=0}^{7} f(x,y) \cos \frac{(2x+1)u\pi}{16} \cos \frac{(2y+1)v\pi}{16} \right] \quad (3.12)$$

Typically, most of the spatial frequencies have zero or near-zero values and can be ignored in the encoding. This fact constitutes the major factor of compression.

The output from the FDCT is quantized according to a table that defines 255 steps into which the DCT coefficients are grouped. The sizes of these steps are specific to the application or specified by the user. Ideally, the step sizes should be selected as close as possible to the thresholds of perceived differences in the corresponding image. Psychovisual experiments can be and have been conducted to establish the preferred thresholds. These values vary with each application. As DCT coefficients with different values are grouped into single steps, the process introduces losses of picture information at the same time as it adds compression of data.

The DCT coefficients obtained can be depicted in an 8 × 8 array, as shown in Fig. 3.18. This arrangement does *not* correspond to the positions of the original picture elements. Of the 64 coefficients, the first

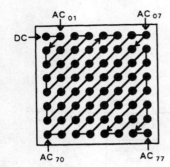

Figure 3.18 Coding array of DCT coefficients.

(0,0) is the DC coefficient, and the other 63 coefficients are of the AC type. Typically, most of the information is contained on the DC coefficient. The coding of the array is performed in a zigzag pattern, as shown in Fig. 3.18. Rather than coding the absolute value of the DC coefficient, the difference between the values of the DC coefficients between adjacent blocks, as shown in Fig. 3.19, is coded and transmitted as per Eq. (3.13):

$$\Delta DC_i = DC_i - DC_{i-1} \tag{3.13}$$

In the last step of the encoder, the quantized DCT coefficients are subjected to an entropy coding method, either of the Huffman or arithmetic type. Huffman coding is prescribed for the baseline sequential codec. However, both Huffman and arithmetic coding are specified as acceptable alternatives for each of the four coding modes. As we know, these are statistical coding formats that further compress the image signals.

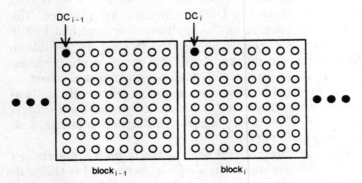

Figure 3.19 DC coding.

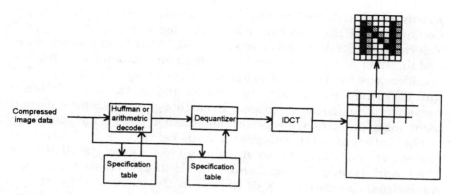

IDCT = inverse discrete cosine transform

Figure 3.20   DCT decoder.

A decoder of the type shown in Fig. 3.20 performs the decoding of the received signals. In the first step, an inverse Huffman or arithmetic coding is performed to reestablish the quantized DCT coefficients. This is based on the Huffman/arithmetic coding tables from the encoder that have been made available at the decoder. In the next step, the coefficients are dequantized, again according to specification tables transmitted from the encoder. In the last step, the signals are subjected to inverse discrete cosine transform (IDCT) and closely converted back to the original signals. As we have seen, they are not identical to the original signals because the quantizing process is not lossless. Based on these signals, an image that is almost identical to the original is reconstructed, as shown in Fig. 3.20.

3.4.6   Moving picture coding

The term *moving pictures* refers to a sequence of pictures that are transmitted or stored one after another. Examples include television, video, and movies. In these media, new or modified picture frames occur at a rate of 25 per second and more. Thus the already high number of bits to represent a still picture has to be multiplied by the number of frames per unit of time to arrive at the total requirement of memory and/or transmission bandwidth. The need for compression is thus higher than in the case of still pictures.

Besides employing the same or similar compression techniques as used for still pictures, the compression of moving pictures takes into account the fact that the changes from one frame to the next usually are small. Several international standards relating to moving picture coding and compression have been issued over the years.

Recommendation 601 of the International Radio Consultative Committee (CCIR) with the title, "Encoding Parameters of Digital Television for Studios," was issued in 1980. It has been revised twice, and the 1990 version is referred to as Recommendation 601-2. It gives specifications for the digital coding of television signals with both 525 and 625 lines and applies to studios and to the international exchange of television programs. Even though the recommendation mentions quantization, no specific quantization table is included.

The International Telegraph and Telephone Consultative Committee (CCITT) has been studying a video coding standard since 1984 and issued Recommendation H.261, "Video Codec for Audiovisual Services at $p \times 64$ kbit/s," in December of 1990. It provides for the transmission of video signals over a group of 64-kbit/s digital channels, typically integrated services digital network (ISDN) channels. The number of channels p can vary from 1 to 30, resulting in bandwidths from 64 to 1920 kbit/s. The T1 committee of the American National Standards Institute (ANSI) has issued a different version of this standard, T1.314-1991, entitled, "Digital Processing of Video Signals—Video Coder/Decoder for Audiovisual Services at 56 to 1536 kbit/s." It takes into account the fact that the bandwidth of each channel in many T1 carrier systems is still limited to 56 kbit/s and specifies multiples of 56-kbit/s channels for video transmission. The upper limits refer to the useful bandwidth of the European E1 carrier system (1920 out of 2048 kbit/s) and the North American (and Japanese) T1 system (with 1536 out of 1544 kbit/s).

The noninterlaced video pictures are coded as luminance and two color difference components called Y, C_B, and C_R. These values are obtained from the primary analog signals E'_R, E'_G, and E'_B as follows: First, the analog signals are digitized to arrive at the values E'_{R_D}, E'_{G_D}, and E'_{B_D} according to the formulas

$$E'_{R_D} = \text{integer } (219E'_R) + 16 \tag{3.14a}$$

$$E'_{G_D} = \text{integer } (219E'_G) + 16 \tag{3.14b}$$

$$E'_{B_D} = \text{integer } (219E'_B) + 16 \tag{3.14c}$$

From these, the values of Y, C_B, and C_R are obtained as follows:

$$Y = \frac{77}{256} E'_{R_D} + \frac{150}{256} E'_{G_D} + \frac{29}{256} E'_{B_D} \tag{3.15a}$$

$$C_R = \frac{131}{256} E'_{R_D} - \frac{110}{256} E'_{G_D} - \frac{21}{256} + 128 \tag{3.15b}$$

$$C_B = -\frac{44}{256}\,E'_{R_D} - \frac{87}{256}\,E'_{G_D} + \frac{131}{256}\,E'_{B_D} + 128 \qquad (3.15c)$$

Two different formats are recommended for the sampling of the analog signals. One is called *common intermediate format* (CIF), in which the luminance Y of a picture is sampled at a resolution of 352 pels per line and 288 lines per picture. The two color difference components C_R and C_B are sampled at 176 pels per line and 144 lines per picture.

Four luminance blocks Y with 8×8 pels each combined with two chrominance blocks C_R and C_B form what is called a *macroblock*, as shown in Fig. 3.21. Thirty-three macroblocks arranged in three rows with 11 blocks in each row form a *group of blocks*, as shown in Fig. 3.22. That figure also shows the arrangement of 12 groups of blocks in two columns that constitute a common intermediate format (CIF) picture.

An alternative format called *quarter common intermediate format* (QCIF) is based on one-quarter of the CIF samples. A QCIF picture consists of three groups of 33 macroblocks arranged in one single column.

Recommendation H.261, cited above, calls for a picture frame frequency of 30,000/1001 per second, or about 29.97 frames/s. At this frame rate, the transmission of uncompressed signals according to CIF requires a bandwidth of 36.45 Mbit/s, and QCIF requires 9.115 Mbit/s. Significant amounts of compression are required in order to transmit these signals over twenty-four 56-kbit/s channels or less. A

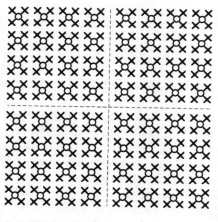

✕ — Luminance
○ — Chrominance

Figure 3.21 Luminance and chrominance samples in a macroblock.

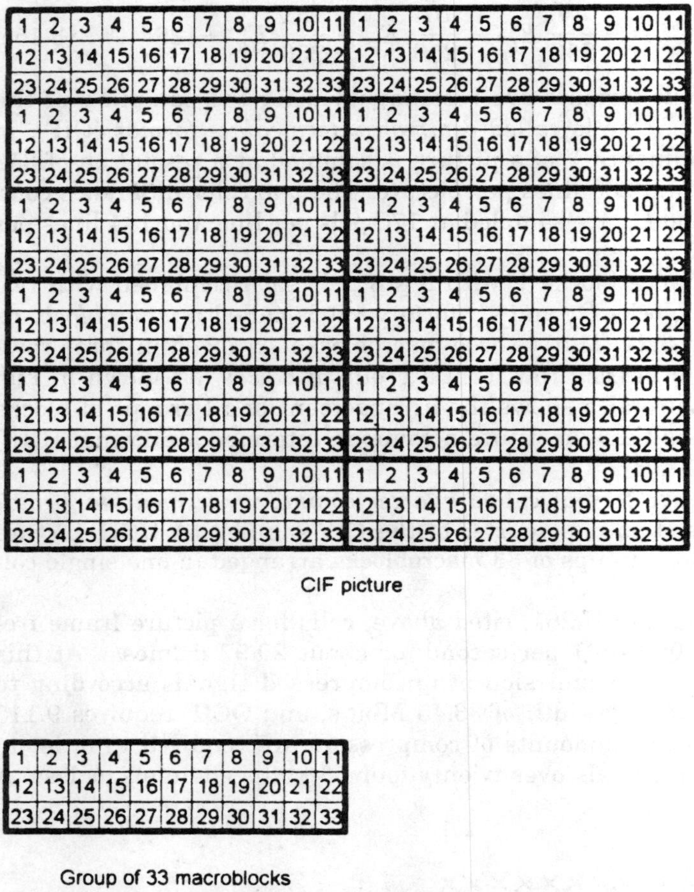

1	2	3	4	5	6	7	8	9	10	11	1	2	3	4	5	6	7	8	9	10	11
12	13	14	15	16	17	18	19	20	21	22	12	13	14	15	16	17	18	19	20	21	22
23	24	25	26	27	28	29	30	31	32	33	23	24	25	26	27	28	29	30	31	32	33
1	2	3	4	5	6	7	8	9	10	11	1	2	3	4	5	6	7	8	9	10	11
12	13	14	15	16	17	18	19	20	21	22	12	13	14	15	16	17	18	19	20	21	22
23	24	25	26	27	28	29	30	31	32	33	23	24	25	26	27	28	29	30	31	32	33
1	2	3	4	5	6	7	8	9	10	11	1	2	3	4	5	6	7	8	9	10	11
12	13	14	15	16	17	18	19	20	21	22	12	13	14	15	16	17	18	19	20	21	22
23	24	25	26	27	28	29	30	31	32	33	23	24	25	26	27	28	29	30	31	32	33
1	2	3	4	5	6	7	8	9	10	11	1	2	3	4	5	6	7	8	9	10	11
12	13	14	15	16	17	18	19	20	21	22	12	13	14	15	16	17	18	19	20	21	22
23	24	25	26	27	28	29	30	31	32	33	23	24	25	26	27	28	29	30	31	32	33
1	2	3	4	5	6	7	8	9	10	11	1	2	3	4	5	6	7	8	9	10	11
12	13	14	15	16	17	18	19	20	21	22	12	13	14	15	16	17	18	19	20	21	22
23	24	25	26	27	28	29	30	31	32	33	23	24	25	26	27	28	29	30	31	32	33
1	2	3	4	5	6	7	8	9	10	11	1	2	3	4	5	6	7	8	9	10	11
12	13	14	15	16	17	18	19	20	21	22	12	13	14	15	16	17	18	19	20	21	22
23	24	25	26	27	28	29	30	31	32	33	23	24	25	26	27	28	29	30	31	32	33

CIF picture

1	2	3	4	5	6	7	8	9	10	11
12	13	14	15	16	17	18	19	20	21	22
23	24	25	26	27	28	29	30	31	32	33

Group of 33 macroblocks

Figure 3.22   Arrangement of groups in a CIF picture.

36.46-Mbit/s bit stream has to be compressed about 1:30 to fit on a 1344-kbit/s channel.

Figure 3.23 shows a block diagram of a typical H.261 video codec. An analog (or digital) video signal enters a source coder from the top left of the figure. There it is subjected to either intraframe or interframe coding under a coding control mechanism. *Interframe coding* is used for sequences of similar pictures, while *intraframe coding* is used for the first picture of a new sequence. Interframe coding removes redundancies between consecutive pictures in a sequence. In the box called "transform," the signals are subjected to a hybrid of discrete cosine transforms (DCT) and, in interframe mode, differen-

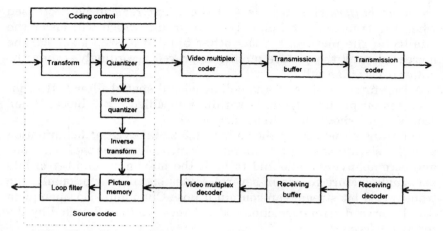

Figure 3.23 Block diagram of H.261 video codec.

tial pulse code modulation (DPCM). Each individual 8×8 block of pels is transformed into DCT coefficients and linearly quantized in the quantizer. The order in which transform coefficients are transmitted is the same as the one shown for the JPEG standard in Fig. 3.18. The quantized coefficients are then sent both to the video multiplex coder and toward the picture memory in the decoding branch for use in connection with interframe coding. On the way to the picture memory they are subjected to inverse quantizing and inverse transform. In the interframe mode, the luminance values Y of the current macroblock are compared with those of neighboring macroblocks in the previous picture. If the difference is below a certain value, no information is transmitted. If there is a significant difference, the difference values are transformed according to DCT, linearly quantized, and transmitted to the video multiplex coder.

The *video multiplex* consists of a hierarchical structure with six layers: video sequence, group of pictures, picture, slice, macroblock, and block. The coding control mechanism generates overhead information for each layer that is multiplexed with the output from the quantizer. The overhead includes quantizing tables, coding tables, etc. Information regarding each layer is entropy coded and transmitted in the form of *variable-length code words* according to coding tables of the Huffman type that are part of the recommendation and standard, respectively. The number of bits transmitted for each picture can vary between a large number for intraframe coding and zero bits for skipped pictures. Thus the transmitted bit stream is buffered in a *transmission buffer,* and the received bit stream is buffered in a *receiving buffer.* The step sizes of the quantizer can be decreased when the transmission buffer is close to full, which will lower the quality of the picture. On the other hand, the step sizes can be increased—and the picture quality improved accordingly—when the buffer is less than full.

A loop filter consists of a two-dimensional spatial filter (FIL) that operates on picture elements within a specified 8×8 block. When required, it removes high-frequency noise.

The transmission coder shown in Fig. 3.23 takes care of elimination as well as stuffing of bits to maintain synchronization between video and accompanying audio and to limit the maximum number of bits per single picture to one of several values specified. Optionally, the transmitted bit stream can contain a Bose-Chaudhuri-Hocquenquem (BCH) forward error-correction code. Errors are then detected by the receiving decoder.

CCITT Recommendation H.261 and ANSI standard T1.314-1991 are mainly suitable for use with video telephony and similar applications where reduced quality of the received video pictures can be accepted. To meet the demand for higher-quality video pictures, the Moving Picture Experts Group (MPEG) has been working on a standard for video coding at bandwidths of 1.5 Mbit/s and higher since 1988. The ISO in combination with CCITT now acts as an umbrella for these efforts. A first version, MPEG-1, was approved by the ISO in November of 1992 as ISO/IEC 11172. Work on MPEG-2 is currently under way. Twelve different picture formats are under consideration for MPEG-2, and the standards body cannot decide on a single one. Rather, many will be offered. Table 3.20 shows the ones under final discussion. They are divided into three profiles and four levels of picture quality. The high quality level corresponds to that of high-definition television (HDTV), while the normal quality level corresponds to that of today's standard television sets. The simple and main profiles of the high quality level are intended for HDTV in the United States, while the three highest quality levels of the "next profile" are intended for the European type of HDTV. A broad range of manufacturers of television equipment are willing to accept the normal quality level of the main profile. The high quality level is divided into one with 1920 pel/line, preferred by the Americans, and one with 1440 pel/line, preferred by the Europeans. It also should be noted that the frame frequency has been disassociated from the power-line frequency and that countries outside North America (which have a power frequency of 50 Hz) accept a frame frequency of 60 frame/s. Nobody has shown any interest in developing a standard for the low quality level of the "next" profile.

TABLE 3.20   Proposed MPEG-2 Standard

		Simple profile	Main profile	Next profile
High level	pel/line	1920	1920	1920†
	Mbit/s	<60	<60	<60
	Lines/frame	1152	1152	1152
	Frame/s	60	60	60
	pel/s	62.7×10^6	62.7×10^6	62.7×10^6
High level	pel/line	1440	1440	1440
	Mbit/s	<60	<60	<60
	Lines/frame	1152	1152	1152
	Frame/s	60	60	60
	pel/s	47.0×10^6	47.0×10^6	47.0×10^6
Normal level	pel/line	720	720*	720
	Mbit/s	<15	<15	<15
	Lines/frame	576	576	576
	Frame/s	30	30	60
	pel/s	10.4×10^6	10.4×10^6	11.06×10^6
Low level	pel/line	352	352	352
	Mbit/s	<4	<4	<4
	Lines/frame	288	288	To be decided
	Frame/s	30	30	To be decided
	pel/s	2.53×10^6	2.53×10^6	To be decided

*The solid box represents most of the industry.
†The dashed box represents European HDTV. Balance of data represents U.S. HDTV.

The goal is to incorporate the following features in the MPEG standard:

1. Combination and synchronization of audio and video information
2. Small delays (less than 150 ms) in coding and decoding, including encoding in real time
3. Accommodation of standard tape features, such as
   a. Random access to stored information
   b. Fast forward/reverse searches
   c. Reverse playback
4. Robustness to bit errors in storage and transmission
5. Possibility to edit audio and video information in compressed form
6. Algorithms that can be implemented economically on integrated circuits

The compression algorithm used by MPEG is similar to the ones used by JPEG and CCITT in Recommendation H.261/ANSI T1.314.

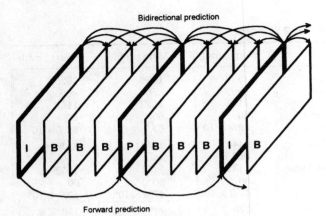

Figure 3.24   Frame sequence.

Thus MPEG employs intraframe and interframe coding, discrete cosine transform, and quantizing. The procedures to compensate for motion between pictures are based on basic blocks consisting of 16 × 16 picture elements. Two different *interframe* coding techniques are used: predictive and interpolative. Either of these techniques leads to so-called *predicted frames* (P). In addition, we have *intraframes* (I) and *bidirectionally predicted frames* (B), as shown in Fig. 3.24. As in the case of the previously discussed coding standards, *intraframe* refers to the first picture of a new sequence, and *interframe* refers to one of a sequence of similar pictures. Predicted frames are coded with reference to a past frame, which can be an intraframe or another predicted frame. Bidirectional frames use both past and predicted frames as references. Bidirectional prediction, also called *motion-compensated interpolation,* contributes significantly to a high image quality; a high compression decreases noise and error propagation.

Motion-compensated interpolation predicts a new 16 × 16 block macroblock of a picture from an older one—forward prediction—or by a prediction based on earlier and future pictures—bidirectional prediction. Specifically, several motion vectors are calculated and used. Among them are vectors for forward predicted macroblocks and backward predicted macroblocks, as well as two vectors for bidirectionally predicted macroblocks.

Arriving at the final standard involved several compromises. The requirement for random access to stored video information is best accomplished through the use of intraframe coding alone. On the other hand, the desired quality requirements cannot be met with intraframe coding alone but need a balance with interframe coding.

The selection of the two interframe coding techniques, predictive and interpolative, is also a means of accomplishing a high image quality.

A committee within CCIR, called CMTT, for *Commission Mixte des Transmission Télévisuelles et Sonores,* is working on CCIR recommendations for the transmission of television signals and associated sound. A special task group within CMTT, called CMTT/2, is dealing with recommendations for the transmission of television signals, both conventional and high-definition television (HDTV), for primary distribution, for secondary distribution, and for contribution. As defined by CMTT, *contribution* concerns a link, for example, between production centers or studios that are designed to allow for appropriate postproduction. *Primary distribution* is between a production center (or a studio) and, for example, a cable head, a transmitter, or a satellite earth station. *Secondary distribution* refers to the mass distribution with preference to broadband ISDN but also over satellite, CATV, and terrestrial networks. In the case of primary distribution and contribution, CMTT/2 is working on CCIR Recommendation 723 for television transmission at the bit rates of 34 and 45 Mbit/s and on Recommendation 721 for transmission at 140 Mbit/s. A Special Rapporteur to Task Group CMTT/2 is charged with the studies for coding and transmission and is working closely with CCITT-MPEG in order to produce a common standard.

3.5   Protocols

3.5.1   General

A data communications *protocol* is a set of conventions regarding the timing, format, and content of data to be exchanged among terminals and computers. Among the basic functions of protocols are

1. Synchronization of bits
2. Synchronization of bytes
3. Synchronization of messages and packets
4. Positive acknowledgement of reception (ACK)
5. Negative acknowledgment of reception (NAK)
6. Error detection
7. Error correction

In order to put order in the many different protocols available and in use, the ISO introduced the open systems interconnection (OSI) seven-layer model shown in Table 3.21. Each layer depends on the underlying layers with lower numbers and supports upper layers

TABLE 3.21   The ISO-OSI Model

7. Application
6. Presentation
5. Session
4. Transport
3. Network
2. Data link
1. Physical

with higher numbers. As the name indicates, the lowest layer deals with the physical—i.e., electrical, mechanical, functional, and procedural—characteristics of an interconnection of links in a data communications network. Layer 2, the data link, is responsible for the transport and error-free delivery of data over the physical link—in particular, error detection, error recovery, and sequence control. The network layer, layer 3, controls the establishing, maintaining, and releasing of connections. It performs switching and selects the proper route between terminals. Layer 4, the transport layer, is responsible for the transport of data for the higher layers over the lower layers. Layer 5 establishes sessions between users. The presentation layer, layer 6, is responsible for assembly of the data and ensures that the data are properly presented to the application layer, which supports the users' application programs. As data communication and its protocols become more complicated, it has proven necessary to split one or several OSI layers in two or more in certain cases and to create sublayers. Even though the OSI model was originally designed for data transmission, it is often applied to voice, image, and video transmission as well.

3.5.2 Asynchronous transfer mode (ATM) protocol and cells

The CCITT (now the ITU-T) recommends that broadband integrated services digital networks (B-ISDNs) be implemented in asynchronous transfer mode (ATM). This fact and the fact that ATM has grown to become very popular are strong indications of a bright future for a combined B-ISDN and ATM.

In the ATM environment, each user negotiates a contract with the B-ISDN (network) that specifies class or type of service that the user requires. These four classes (A, B, C, and D) are discussed below and shown in Fig. 3.25. Usually for each call the user also negotiates and contracts for a specific quality of service that may include traffic delays and rates of cell losses. The user also should indicate the char-

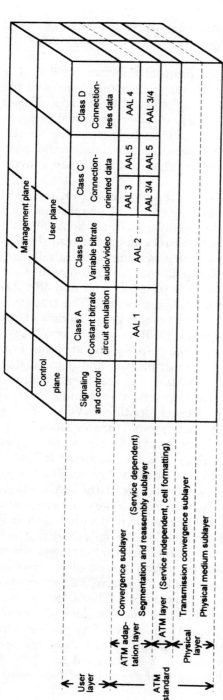

Figure 3.25 B-ISDN/ATM protocol reference model.

acteristics of the traffic to be transmitted with regard to the peak cell rate and average cell rate. In other words, the user negotiates the bandwidth of the connection.

Figure 3.25 shows a B-ISDN/ATM protocol reference model. It is divided into three planes: the user plane, the control plane, and the management plane. The user plane handles the flow transfer of user information and associated controls, such as flow control and recovery from errors. Functions in the control plane include signaling for call setup, maintenance, and release. A separate plane handles the management functions. The user and control planes have a layered structure.

At the bottom of the stack of layers is the physical layer, on top of which comes the ATM layer, followed by the ATM adaptation layer (AAL). These three layers constitute the layers within the ATM standard. The physical layer is split into two sublayers: the physical medium sublayer and the transmission convergence sublayer. At all edges of the ATM system, information is transferred in fixed-length packets, called *cells*. A cell consists of a 5-byte (octal) header and 48 bytes (octals) of user payload for a total of 53 bytes (octals). The transmission convergence sublayer transforms a flow of cells into a flow of bits that can be transmitted over the physical medium and the reverse.

Functions in the ATM layer format the cells for transmission over the physical layer. This includes cell multiplexing and demultiplexing, translations of virtual path and virtual channel identifiers, cell header generation and extraction, and generic flow control. Virtual path identifiers (VPIs) and virtual channel identifiers (VCIs) are discussed in Sec. 10.3, "Switching in Asynchronous Transfer Mode."

The ATM adaptation layer (AAL) performs the functions related to the adaptation of the users' requirements, as expressed in higher layers, to the functions of the ATM and physical layers. The higher layers can require a wide range of services, and the AAL adapts these services to the ATM layer. Thus the AAL is service dependent. The different services of the higher layers are divided into four classes called class A, B, C, and D based on the following parameters: (1) timing relation between source and destination (required or not required), (2) bit rate (constant or variable), and (3) connection mode (connection-oriented or connectionless). Figure 3.26 shows the relationships between classes and parameters. There are several types of AAL protocols, each designed for a specific class of service. The type 1 AAL is designed for class A, which requires timing relation between source and destination, constant bit rate, and connection-oriented mode. Similarly, AAL type 2 is designed for class B, which differs from class A by allowing for a variable bit rate rather than a fixed one.

Class of service	Class A	Class B	Class C	Class D
Timing relation between source and destination	Required		Not required	
Bitrate	Constant	Variable		
Connection mode	Connection-oriented			Connectionless
AAL	AAL 1	AAL 2	AAL 3 / AAL 5 / AAL 3/4	AAL 4 / AAL 5 / AAL 3/4

Figure 3.26 ATM adaptation layer.

For classes C and D, the AAL is divided into two sublayers: a segmentation and reassembly sublayer and a convergence sublayer. The segmentation sublayer breaks up higher-level information into ATM cells sizes, and reassembly reverses this function. The convergence sublayer accommodates the different bit streams of the higher layers to the ATM layer. AAL type 3 handles connection-oriented data service and signaling, while AAL type 4 supports connectionless data services. While AAL type 3 and AAL type 4 differ in the convergence sublayer, they are identical in the segmentation and reassembly sublayer.

AAL type 5 is a new variety of AAL types 3 and 4 with a simplified structure. (It was originally called *simple and efficient AAL,* or SEAL.) It performs a subset of the functions of AAL types 3 and 4. AAL type 5 handles message mode service only and not streaming mode service. The difference between these services is that in the message service mode, service data units (SDUs) passing across the ALL type 5 common part interface are identical to the interface data units (IDUs). In the streaming mode, the data units (SDUs) passing across the AAL type 5 common part interface consist of one or more interface data units (IDUs).

A protocol data unit (PDU) can be 65,536 bytes long and has one overhead. When converted to AAL types 3 and 4, the overhead is repeated in each cell, while in AAL type 5 there are one header and one trailer for the entire PDU.

3.6 Packets

In certain applications, data are packetized and transmitted and switched in the form of packets. A packet with variable size is also called a *frame,* while a packet of a fixed size is called a *cell.* Each packet contains an address that indicates the addressee of the information and sometimes the desired routing of the packet as well as identification of the sender. This information is contained in a header. Sometimes a packet has a trailer, which typically contains data used for error detection and error correction.

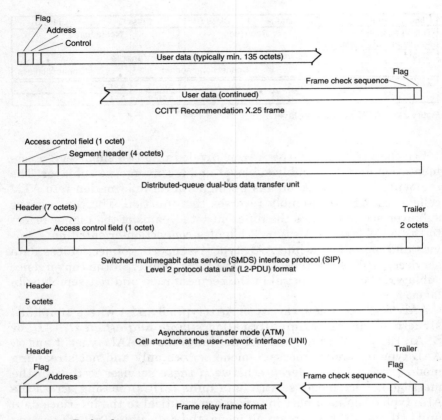

Figure 3.27 Packet formats.

Figure 3.27 shows the formats of some packets, and Table 3.22 lists some comparisons between packet switching systems. The packets (or frames) recommended by CCITT in Recommendation X.25 are included as a reference, as is frame relay, even though none of them can support broadband transmissions. Actually, only ATM can support broadband transmissions of all kinds.

TABLE 3.22 Comparisons of Packet Formats

	X.25	Frame relay	SMDS	ATM
Transmission speed	<56 kbit/s	<1.5 Mbit/s	<44 Mbit/s	>45 Mbit/s
Packet/frame size	Variable	Variable	53 octets	53 octets
Header plus trailer	6 octets	7 octets	9 octets	5 octets
Payload	Variable	Variable	44 octets	48 octets
Normal type of packet switching	Connectionless	Connectionless	Connectionless	Connection-oriented

Frames with variable sizes of packets and payload, such as X.25, and frame relay are more economical with regard to the relation of overhead to payload. On the other hand, packets with a fixed size (cells) can operate at higher speeds, and the switching equipment is less expensive.

References

Standards

IEEE Standard 100 (1977), *IEEE Standard Dictionary of Electrical and Electronics Terms,* IEEE, New York.

CCITT Recommendation G.721 (1989), "32 kbit/s adaptive differential pulse code modulation (ADPCM)," in *Blue Book,* fascicle III.4, CCITT, Geneva.

CCIR Recommendation 601-2 (1990), "Encoding parameters of digital television for studios," in *Recommendations and Reports of the CCIR,* 17th Plenary Assembly, vol. XI-1, part 1, International Telecommunications Union, Geneva, pp. 95–104.

CCITT Recommendation H.261 (1990), "Line transmission on non-telephone signals—video codec for audiovisual services at $p \times 64$ kbit/s," CCITT, Geneva.

ANSI Standard T1.314-1991 (1991), "Digital processing of video signals—video coder/decoder for audiovisual services at 56 to 1536 kbit/s," ANSI, New York.

CCITT Recommendation I.321 (1991), "Integrated Services Digital Network (ISDN)—overall network aspects and functions, ISDN user-network interfaces—B-ISDN protocol reference model and its applications," CCITT, Geneva.

CCITT Recommendation I.361 (1991), "Integrated Services Digital Network (ISDN)—overall network aspects and functions, ISDN user-network interfaces—B-ISDN ATM layer specification," CCITT, Geneva.

CCITT Recommendation I.362 (1991), "Integrated Services Digital Network (ISDN)—overall network aspects and functions, ISDN user-network interfaces—B-ISDN ATM adaptation layer (AAL) functional description," CCITT, Geneva.

ISO International Standard IS-10918-1 (1992), "Digital compression and coding of continuous-tone still images," part 1: "Requirements and guidelines," ISO, Geneva.

ISO International Standard IS-10918-2 (1992), "Digital compression and coding of continuous-tone still images," part 2: "Compliance testing," ISO, Geneva.

Other references

Ahmed, Nasir, T. Natarajan, and K. R. Rao (1974), "Discrete Cosine Transform," *IEEE Transactions on Computers,* vol. C-23, January, pp. 90–93.

Amsterdam, Jonathan (1986), "Data Compression with Huffman Coding," *Byte,* vol. 11, no. 5, May, pp. 99–108.

Berger, T. (1971), *Rate Distortion Theory,* Prentice Hall, Englewood Cliffs, New Jersey.

Huffman, David A., (1952), "A Method for the Construction of Minimum Redundancy Codes," *Proceedings of the IRE,* vol. 40, no. 9, September, pp. 1098–1101.

Kawarasaki, Masatoshi, and Bijan Jabbari (1991), "B-ISDN Architecture and Protocol," *IEEE Journal on Selected Areas in Communications,* vol. 9, no. 9, December, pp. 1405–1415.

Langdon, Glen G., Jr. (1984), "An Introduction to Arithmetic Coding," *IBM Journal of Research and Development,* vol. 28, no. 2, March, pp. 135–149.

Le Gall, Didier (1991), "MPEG: A Video Compression Standard for Multimedia Applications," *Communications of the ACM,* vol. 34, no. 4, April, pp. 46–58.

Lindberg, Bertil C. (1991), "Switched Broadband Network Interfaces," *Telecommunications* (North American Edition), vol. 25, no. 10, October, pp. 47–51, 53, 58.

Mermelstein, Paul (1988), "G.722: A New CCITT Coding Standard for Digital Transmission of Wideband Audio Signals," *IEEE Communications Magazine,* vol. 26, no. 1, January, pp. 8–15.

Nyquist, H. (1928), "Certain Topics in Telegraph Transmission Theory," *Transactions of the AIEE,* vol. 47, April, pp. 617–644.

Pennebaker, William B., and Joan L. Mitchell (1992), *JPEG—Still Image Data Compression Standard,* Van Nostrand Reinhold, New York.

Schwartz, Mischa (1990), "Lightwave-Based, High-Speed Networks of the Future," in *Proceedings of the 80th Communications Society—New York Section seminar on Emerging Technologies for High-Speed Digital Communications—New Architectures and Applications,* Nov. 15.

Wallace, Gregory K. (1991), "The JPEG Still Picture Compression Standard," *Communications of the ACM,* vol. 34, no. 4, pp. 30–44.

Digital Broadband
Network Technology

Digital Broadband Subscriber Loops

4.1 Introduction

The original definition of a *subscriber loop* is the twisted pair of copper wires between a customer or subscriber and the telephone switch serving that customer. Similarly, cable television operators install coaxial cables or fiberoptic cables to each of their customers. Other forms of links between users and service providers include radio links, optical links, waveguides, etc.

4.2 Copper Loops

Over 94 percent of U.S. homes and all businesses are connected to the worldwide telephone network by means of twisted-pair copper wires. The bit-rate carrying capability of these wires depends on the length of the loop, the type of wire used, and the presence of loading coils and bridge taps. Figure 4.1 shows the cumulative percentage of loops by their length. For instance, 40 percent of loops are shorter than 3.2 km. Figure 4.2 shows the carrying capacity in megabits per second by wire gauge and by length of the loop. The layout of a typical analog loop with twisted wires is shown in Fig. 4.3. It includes two hybrids to accommodate two-way traffic on a single pair of wires. A *tap* is basically a branch that has been left from an earlier use of the wire pair, as shown in Fig. 4.4. Such taps interfere with high-frequency traffic,

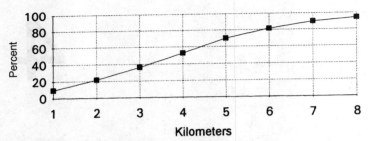

Figure 4.1 Lengths of U.S. subscriber loops.

as do loading coils, i.e., serial coils inserted in the wires to compensate for the capacity of the loop.

The capacity of subscriber loops can be increased so that two pairs of twisted wires can carry 96 channels through the use of *subscriber loop carrier* (SLC) *equipment,* as shown in Fig. 4.5. Even though the SLCs are digital and typically operate at 1.5 Mbit/s, they cannot be used as a 1.5-Mbit/s channel. Theoretically, but hardly practically, the 96 digital channels can be reversed multiplexed to form a single 1.5-Mbit/s channel.

As seen in Fig. 4.2, the capability of carrying traffic on subscriber loops is limited to about 1.5 Mbit/s. Among the limiting factors are near-end crosstalk (NEXT) when traffic with high bit rates is sent in both directions on wire pairs that are in the same cable binder. To some extent, the NEXT problem can be alleviate with decreased repeater spacings.

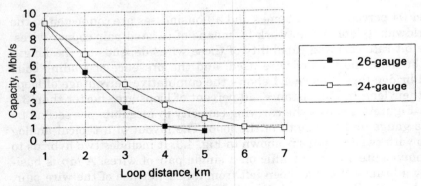

Figure 4.2 Bit-rate capacity of subscriber loops.

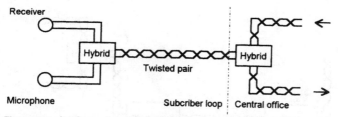

Figure 4.3 Analog subscriber loop with twisted pair.

4.3 Advanced Copper-Based Subscriber Loops

The increasing demand for broadband facilities has resulted in efforts to extend the use of the existing copper-based subscriber loops beyond the bit rates shown in Fig. 4.2. Among technologies being tested and/or implemented are advanced modem technologies, echo cancellation, advanced line coding schemes, and digital signal processing (DSP).

Among techniques being promoted are *high-bit-rate digital subscriber line* (HDSL), *very high-bit-rate digital subscriber line* (VHDSL), *asymmetric digital subscriber line* (ADSL), and *very high-bit-rate asymmetric digital subscriber line* (VHADSL) technologies. T. Russell Hsing et al. (1993) describe HDSL and ADSL.

HDSL requires two unloaded twisted copper pairs and is limited to the so-called carrier serving area (CSA), i.e., within 3 km using 26-gauge and 4 km using 24-gauge wire. ADSL can transmit data at 6 Mbit/s unidirectionally to the customer and offer two-way POTS (4-kHz) or basic ISDN service ($2 \times 64 + 16$ kbit/s) in both directions on a single pair of copper wires.

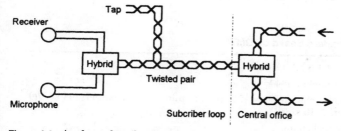

Figure 4.4 Analog subscriber loop with tap.

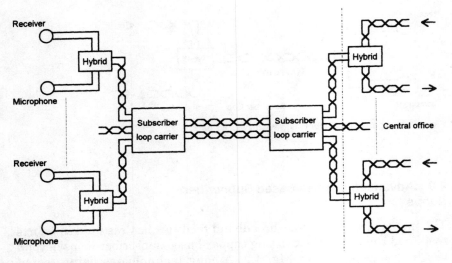

Figure 4.5 Subscriber loop carrier.

An asymmetric digital subscriber line (ADSL) is asymmetric in the sense that it provides broadband transmission toward the subscriber and narrowband transmission away from the subscriber. An original concept, later called ADSL I, allowed for transmission of data up to a rate of 1.5 Mbit/s. A later version, called ADSL II, extended the bit rate to about 6 Mbit/s. Today, the two are combined, and no distinction is apparent. The T1E1.4 subworking group of the U.S. Exchange Carriers Standards Association (ECSA) endorsed DMT as the preferred line coding technique for ADSL on March 10, 1993. This opens the way for the American National Standards Institute (ANSI) to adopt it as a U.S. standard. This ADSL standard allows for transmission of the types of applications listed in Table 4.1 over 4 km of a 24-gauge twisted copper pair. Twenty-six gauge pairs and mixed-gauge pairs will be able to operate over slightly shorter distances.

TABLE 4.1   ADSL Standard Based on DMT

(a)   Four A channels at 1.5 Mbit/s each

(b)   One ISDN H0 channel at 384 kbit/s

(c)   One ISDN basic rate channel (144 kbit/s), containing two B channels (64 kbit/s) and one D channel (16 kbit/s)

(d)   One signaling/control channel

(e)   Basic telephone service operating under these channels

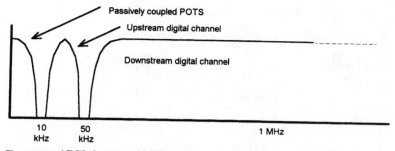

Figure 4.6 ADSL frequency spectrum. (*Source: Amati Communications Corp.*)

Figure 4.6 shows the frequency spectrum of ADSL. Conventional 4-kHz analog telephone service (POTS) is placed below 10 kHz. The POTS is passively coupled to the loop. This means that this part will operate even when the digital part fails. An asymmetric upstream channel with a bit-rate capacity of up to 384 kbit/s is located between 10 and 50 kHz. A broadband asymmetric downstream channel is placed in the frequency spectrum above 50 kHz. The digital channels are based on 256 discrete multitone (DMT) subchannels. As stated earlier, each subchannel has a bandwidth of 4 kHz; i.e., it can carry 4000 symbols/s. Each subchannel with an acceptable performance rating can carry up to 11 bits/symbol. Subchannels with unacceptable characteristics are ignored and not used. The DMT technique was developed by Amati Communications Corporation, a subsidiary of Northern Telecom based in Palo Alto, California. Theoretically, this kind of ADSL system can carry $4000 \times 256 \times 11 = 11,264,000$ bit/s, or about 11.2 Mbit/s. Practically, half this amount, or 6.3 Mbit/s, is the limit. The reason for this is that not all 256 subchannels can carry the full amount of 11 bits/symbol and some subchannels are completely discarded.

Figure 4.7 shows the architecture of a typical ADSL system. Channels of the types listed in Table 4.1 connect a central office ADSL terminal unit to its central office. In addition, an embedded operations channel (EOC) provides OAM&P (operations, administration, maintenance, and provisioning) support, including self-testing, controlled testing, channel analysis, alarms, configuration, and surveillance. Unlike a conventional multicarrier, which would require 256 separate transmitters and receivers, DMT relies on a discrete Fourier transform algorithm called *fast Fourier transform* (FFT) and its inverse (IFFT). A DMT transmitter takes in a block of data, assigns groups of bits to channels based on learned channel charac-

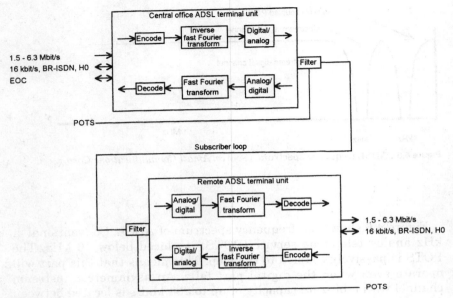

Figure 4.7   ADSL terminal units.

teristics, encodes each channel based on the values of the bits, and passes the ensemble of encoded values to a 512-point IFFT. Computing in the frequency domain, the IFFT generates 256 carrier-less AM/PM subchannels in digital form, with each subchannel capable of conveying from 1 to 11 bits/symbol depending on the measured subchannel signal-to-noise ratio. A standard D/A converter creates an analog signal suitable for coupling to a copper loop. The receiver simply reverses the process.

AT&T Bell Laboratories has developed a technique called *carrierless amplitude and phase* (CAP) *modulation* that lifts the upper transmission bit rates further up. CAP is a passband line coding technique that is a variant of quadrature amplitude modulation (QAM). The signals fit to a passband through filtering as compared with bandpass signals, which fill a needed frequency spectrum.

Figure 4.8 shows possible architectures for CAP transceivers. The incoming data are scrambled by a scrambler in order to eliminate long strings of digital zeros, etc. The customer data then enter an encoder, where they are mapped into complex symbols and further "massaged" by a symbol processing unit. After the encoder, the signals pass through in-phase and quadrature filters. Finally, they are converted to analog signals and sent to the bidirectional subscriber loop via an adaptive hybrid. Signals coming from the loop are convert-

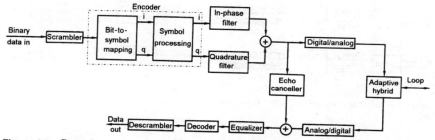

Figure 4.8 Generic architecture for carrierless amplitude/phase (CAP) transceiver. (*Adapted from T1E1.4/90-154 of Sept. 24, 1990, by AT&T.*)

ed to digital signals, which pass an equalizer and then are decoded and descrambled. The system also contains an echo canceler. Several different types of encoders can be used in the CAP system, including some with trellis and other forward error-correcting codes.

AT&T claims that CAP technology can handle bit rates of 20 to 125 Mbit/s over short distances up to 100 m.

4.4   Fiberoptics

As we have just seen, copper wires do not offer sufficient bandwidth for broadband links. The current trend is to substitute fiberoptic cables. Currently, few optical fibers are installed all the way from a central office to the home of a customer, *fiber-to-the-home* (FTTH). Much more often, fibers are installed to businesses. Fibers are used extensively in the feeder network, i.e., from central offices to points close to clusters of customers. From the feeder terminals to the residences and businesses, the term *distribution network* is used. Most distribution cables still consist of twisted-wire pairs. This type of installation is often referred to as *fiber-to-the-curb* (FTTC). The term *fiber-in-the-loop* (FITL) is used with a meaning similar to FTTH.

Optical fibers can carry digital information at rates of up to several tens of billion of bits per second, i.e., tens of gigahertz. When applied in subscriber loops, it is difficult to imagine saturation of a fiber link. It seems difficult to find applications to fill even a fraction of the fibers. Thus the bandwidth is increased by several orders of magnitude compared with older transmission systems. Some time in the future this prediction will be laughed at and compared with one made 100 years ago, when a forecaster said that the market for automobiles in Europe would be limited to 1000 cars or when people around 1945 said that the world could not use more than a half dozen computers.

Fiberoptic links carry information as waves of visual light. They

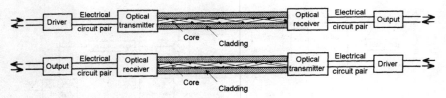

Figure 4.9 Multimode optical fiber transmission link.

are also called *optical fiber links* and *photogenic links*. The light waves are pulse code modulated (PCM), similarly to the modulation used in electrical digital channels, such as T1. At the sending end of a link, electrical pulses are converted into light pulses by semiconductor lasers or light-emitting diodes (LEDs). The links typically have a core of doped silica covered with a cladding with a lower index of refraction. The silica is doped in order to control its index of refraction. The indices of refraction in the core and the cladding are selected in such a way that the light waves are totally reflected at the border between the core and cladding. Thus light waves entering the core at one end will travel down the core. At the receiving end of the link, the light pulses are converted back into electrical pulses by a phototransistor. Arrangements of optical fiber transmission links are shown in Figs. 4.9 and 4.10, with the first showing a multimode fiber and the second a single-mode fiber. The major difference is that a multimode fiber has a large core diameter (>50 μm) in relation to the wavelength of the light, while the core diameter is smaller (<0.85 μm) in a single-mode fiber. Early practical applications used multimode fibers for technical reasons. They are easier to manufacture and to splice. The trend is toward single-mode fibers, which cause less dispersion of the light and thus provides longer links without the need for regeneration.

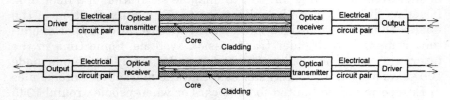

Figure 4.10 Single-mode optical fiber transmission link.

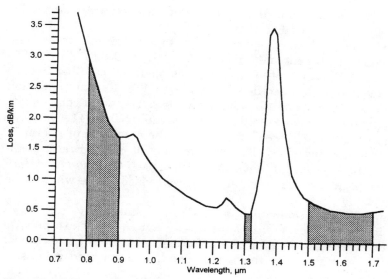

Figure 4.11 Losses and available windows in single-mode optical fibers.

The attenuation of light waves in a fiber depends on the wavelength (or frequency), as shown in Fig. 4.11. The curve shows net spectral loss in decibels per kilometer and includes loss from ultraviolet, infrared, and hydroxyl ion (OH^-) absorption, as well as Rayleigh scattering. The OH^- absorption has peaks at 950, 1250, and 1383 nm. The wavelengths most often used commercially are in the 800–900, 1300–1320 and 1500–1700 nm *windows,* the shadowed areas in Fig. 4.11. The windows represent areas of wavelength where the attenuation is relatively low. From the 800–900 nm window, applications have moved to the 1300–1320 nm window. The window above 1500 nm offers promises for the future, but not all practical problems in that area have been solved yet.

Light waves of different frequencies can be used on the same fiber. One such arrangement is called *wavelength-division-multiplexing* (WDM). Here, several channels are carried on a single fiber at different wavelengths. More than one channel can be inserted within each of the commercially favored windows in Fig. 4.11. A single optical fiber also can carry light waves in two directions by using different wavelengths. WDM and two-way fibers are seldom used because of the relatively low cost of fibers compared with the cost of wavelength-splitting equipment. On long links, where the cost of the terminal equipment is insignificant, the traffic volume is usually high enough to justify more than one fiber and to dedicate separate fibers for transmission in the two directions.

In its technical advisory regarding user-to-network interfaces (subscriber loops) for broadband ISDN at 51, 155, and 622 Mbit/s (TA-NWT-001112), Bellcore specifies one single-mode optical fiber in each direction of the type specified in ITU-T/CCITT Recommendation G.652 and also known as EIA class IVa fiber. Trends in components of optical fiber transmission links are discussed in Chap. 11.

Operating companies are contemplating the extension of fiberoptic links to customer premises. This will make a lot of bandwidth available. Even though some of that bandwidth is used for high-speed data transmission, image transport, and two-way video, there will still be abundant bandwidth left.

Several telephone companies in the United States have conducted fiberoptic trials of providing services by optical fibers to private homes. Some are listed in Table 4.2. Later trials tested the technical feasibility and market for video-on-demand, i.e., switched video ser-

TABLE 4.2   Fiberoptic Trials to Homes

Telephone company	Location of trial	Service started	Number of homes	Type of service
Southern Bell Telephone & Telegraph Co., Atlanta, Ga.	Hunters Creek, Fla.	11/12/86	251	CATV transport
Southern Bell Telephone & Telegraph Co., Atlanta, Ga.	Heathrow, Fla.	6/30/88	4000	POTS, ISDN, CATV transport
New Jersey Bell Telephone Co., Newark, N.J.	Princeton Gate, N.J.	8/9/88	104	POTS, data
Southwestern Bell Telephone Co., St. Louis, Mo.	Leawood, Kan.	10/27/88	134	POTS
South Central Bell Telephone Co., Birmingham, Ala.	Memphis, Tenn.	11/16/88	99	POTS
Contel Corp., Atlanta, Ga.	Ridgecrest, Calif.	12/28/88	400	POTS
Bell of Pennsylvania, Philadelphia, Pa.	Perryopolis, Pa.	7/1/89	100	POTS, CATV transport
U.S. West Communications, Englewood, Colo.	Mendota Heights, Minn.	1Q, 1989	100	POTS
Contel Corp., Atlanta, Ga.	Sidney, N.Y.	2Q, 1989	600	POTS
GTE Service Corp., Stamford, Conn.	Cerritos, Calif.	2Q, 1989	5000	POTS, CATV, broadband services
Southern Bell Telephone & Telegraph Co., Atlanta, Ga.	Coco Plum, Fla.	2Q, 1989	200	POTS
Southern Bell Telephone & Telegraph Co., Atlanta, Ga.	Governors Island, N.C.	2Q, 1989	49	POTS

SOURCE: *IEEE Spectrum*, vol. 26, no. 2, Feb. 1989, p. 47. © 1989 IEEE. Reprinted with permission.

vice available through dial-up. Based on these trials, telephone operating companies have applied to the FCC for permission to offer video-on-demand as a tariffed permanent commercial service. As of March 31, 1994, seven telephone companies had filed applications covering potential service to 3.9 million homes.

Most of these trials were concerned with plain old telephone services (POTS) only. In a few cases, the trials offered ISDN and data transport to residential customers. Also in a few cases, the telephone companies transported cable television (CATV) to customers on behalf of CATV companies. Since telephone companies are normally not allowed to transport CATV signals, special permission from the FCC was required. In most cases, such permits were granted for 1-year trials only. In the long run this raises the question of whether telephone companies should be allowed to use their broadband links to supply cable television to telephone customers.

In most of the trials, fiberoptic cables were wired all the way to each participating household and a box where the light waves were converted into electrical signals. Similarly, outgoing electrical signals were converted to light waves. Typically, two telephone channels were provided. In the case where the telephone company provided transport of television for a CATV company, typically two channels were provided. In some cases, enough bandwidth for over 30 TV channels was provided. A complication with fiberoptics is that it does not carry any electricity (as a conventional POTS subscriber loop does). Thus electrical power must be supplied locally to drive the optical/electrical conversion, the telephone sets, etc. It also means lack of backup power in case of an electrical outage.

As an alternative to installing fiberoptic cables all the way to the residences, such cable was laid to a distribution box near the house, and conventional copper wire and/or coaxial cable was used to reach inside the house. In some cases, subscriber loop carrier systems were used between the telephone company's central office and the distribution box. The installation of fiberoptic cables and equipment at the trial at Perryopolis, Pa., is estimated to have cost Bell of Pennsylvania an average of $5000 per subscriber. This is high compared with the $1200 to $1500 for POTS. However, the fibers provide more channels, and the price of fiber is expected to come down and be competitive with POTS toward the end of the 1990s.

Evaluation of the trials, application for regulatory permits, and granting of permits will delay the introduction of fiberoptic cables to residences. The industry is, however, confident that it will come. At the end of 1991, there were 1542 km of local loops equipped for video. Significant numbers will not be reached until the end of the century.

The French Telecom has wired 1500 houses in Biarritz with fiberoptic links. Commercial service has been open over these links since 1984. In addition to conventional narrowband services such as telephony and the French videotex service, Minitel, the following broadband services are offered: videotelephony, access to audiovisual data bases, television, stereo sound programs, and on-line television program library. In order not to distort the test results, no free service has been given; rather, commercial fees have been charged. Both residential and business customers are served. So far the services have been well received by the customers.

4.5 Cable Television Networks (Coaxial Cables)

Cable television (CATV) networks carry television programs from a head to individual customers. Currently, most such networks use coaxial cables consisting of a center wire surrounded by a conducting tube. The area between the central conductor and the metallic shield can consist of a dielectric material or of washer-like dielectric material along the cable at equidistances. Coaxial cables typically can carry electromagnetic waves at frequencies up to 3 GHz. Figure 4.12 shows the basic layout of a typical system.

Cable companies are installing fiberoptic cables as substitutes for coaxial cables. The main reasons are that fiberoptic cables are less expensive and easier to handle.

In order to transmit information from the customers to the "rest of the world," a two-way system is required. This differs from the—until now—conventional way of transmitting television programs to the customers only. One way of accommodating two-way transmissions is to send information from the customers to the head at low frequencies and information toward the customers at high frequencies. In one application, the *up channel* (from the customer to the head) operates in the frequency band of 5 to 108 MHz and the *down channel* in the band of 160 to 400 MHz. Filters are used to separate the two bands, as shown in Fig. 4.13. The "up" (L) and "down" (H) channels are separated by filters.

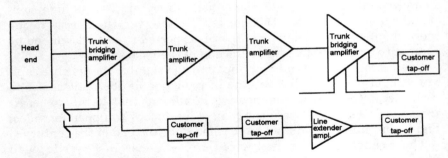

Figure 4.12   Cable television (CATV) network.

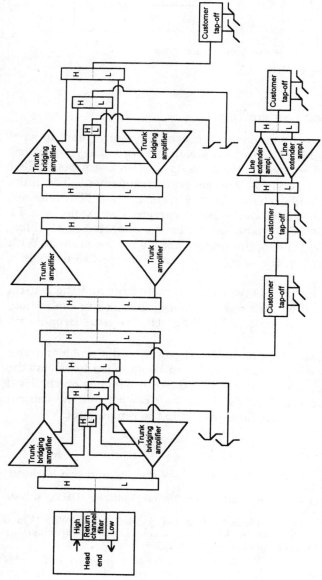

Figure 4.13 Two-way cable television (CATV) network.

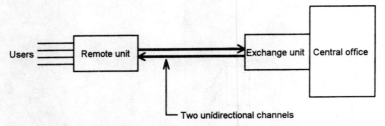

Figure 4.14   Subscriber loop carrier.

4.6   Subscriber Loop Carriers

T1 (and E1) carrier systems have been used for many years to save lines between a cluster of users and a central office. They are called (digital) *subscriber loop carriers* from the days when users were called *subscribers.* Figure 4.14 shows the general arrangement. The links between the remote and exchange units can consist of a T1 carrier, an optical carrier, or a coaxial cable–based carrier. For example, a T1 carrier transports 24 DS-0 channels that can be allocated to 24 different users. Coaxial and fiberoptic carriers handle more channels. With the arrangement shown, the subchannels are arranged to users in a fixed manner.

Further, savings in links between the remote unit and the central office can be obtained by including switching features in the remote unit. This is discussed further in Chap. 10, "Digital Broadband Switching."

Remote units usually require electrical power, and a problem arises regarding how to supply that power. It can be supplied locally at the remote unit, in which case operation of the unit depends on the local power supply. More reliable power can be drawn from the central office, but this requires metallic wires out to the remote unit.

References

Allard, Frederick C. (1990), *Fiber Optics Handbook—For Engineers and Scientists,* McGraw-Hill, New York.
Amati Communications Corp. (1993), "ADSL and Discrete Multitone (DMT)," private communication from Amati.
Hsing, T. Russel, Cheng-Tie Chen, and Jules A. Bellisio (1993), "Video Communications and Services in the Copper Loop," *IEEE Communications Magazine,* vol. 31, no. 1, January, pp. 62–68.

5

Subscriber Interfaces

5.1 Introduction

The term *subscriber loop* refers to the link between a user's terminal equipment and the first switch. This switch is typically a local exchange belonging to a local exchange carrier (LEC). The most common user terminal equipment is a conventional analog telephone set. It also can consist of data terminals, which may be connected via modems (modulator/demodulators), as well as other types of equipment. Technically similar loops also exist between user terminal equipment and private exchanges, such as a private branch exchanges (PBXs).

5.2 Conventional Analog Interfaces

Each type of subscriber loop has a distinct interface. A typical interface handles the setting up of a connection, maintaining of the connection, and the breakdown of the connection. The setting up of a connection involves signaling that a connection is desired (going off-hook), waiting for acknowledgment (dial tone), sending the address of the desired party (dialing), again waiting for acknowledgment (ring or busy tone), and establishing a connection once the called party answers. Basically, the same routine is followed whether the call is a voice, data, or video call. The breaking down of a connection is signaled by one or both parties disconnecting, i.e., going on-hook.

Figure 5.1 shows the interface between a conventional analog telephone set and the local switch. A conventional subscriber loop typical-

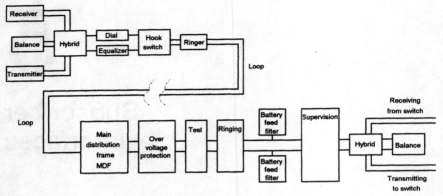

Figure 5.1   Conventional analog telephone set and central office BORSHT interface.

ly terminates in a *line card* at the local switch. In the figure, the line card contains a BORSHT (battery, overvoltage, ringing, supervision, hybrid, and testing) circuit.

Figure 5.2 shows a data terminal interfaced with a subscriber loop through a demodulator and a modulator in the form of a modem. With the use of modems, data traffic with a bit rate of up to 19,200 bit/s can be transmitted over conventional analog telephone circuits. Through the use of modern digital signal processing (DSP) circuits, higher bit rates may be accomplished. In Fig. 5.2, a telephone is connected in parallel to the data terminal.

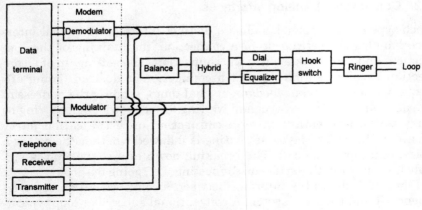

Figure 5.2   Data terminal with modem interface.

5.3   Packet Switching and Frame Relay

Packet switching is one of several techniques for transmitting digital information over a network. The data information is divided into groups, and each group made into a packet. In addition to the payload data, a packet has an address field, which contains information about the destination of the packet and maybe about the route the packet shall take. Packets may take different routes between the source and the final destination, routes that may cause a variety of delays and thus cause the packets to arrive out of order. Thus the address field also should indicate the numerical order of the packets so that they can be sorted and rearranged in the right order at the destination point.

Figure 5.3 depicts a packet switched network with four nodes and two data terminals with their associated packet assembler/disassemblers (PADs) and data circuit-termination equipment (DCE). As the name implies, a packet assembler assembles the incoming data stream into packets, while the packet disassembler disassembles the packets and arranges them back to their earlier input format. Each node reads the address of each packet and directs the packet toward its destination. Depending on the congestion on a node's output links, packets can be routed on different outputs links. Data circuit-termination equipment (DCE) handles the interface between the user equipment and the network. In general terms, a DCE can be an acoustic coupler, modem, data service unit, multiplexor, or concentrator. A data service unit (DSU) is the typical DCE in the arrangement shown in Fig. 5.3.

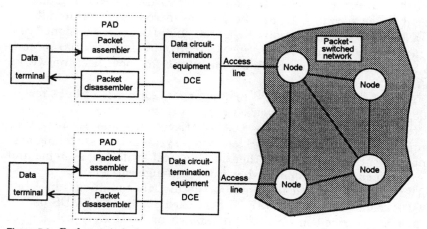

Figure 5.3   Packet-switched network.

The first worldwide standard for packet switching is (ITU-T) CCITT Recommendation X.25. It was designed when the quality of transmission links was questionable. Thus the standard includes error detection on each link between nodes. This slows down the transmission and makes X.25 uneconomical for modern high-quality links. Further, the maximum bit rate for X.25 transmissions is 56 kbit/s, not very much measured against current standards. The X.25 packet format is described in Sec. 3.6 and shown in Fig. 3.27.

In the late 1980s, an improvement called *frame relay* was introduced. The error-detection feature was abandoned and moved to high protocol levels; i.e., it is handled by the two terminals rather than at each node. Further, the nodes relay groups of information at the frame level rather than at the packet level—thus the name *frame relay*.

Frame relay uses variable packet sizes, which is advantageous for data traffic but which makes frame relay unsuitable for voice and video. The frame-relay packet format is described in Sec. 3.6 and shown in Fig. 3.27.

To the extent that frame-relay services are offered to the public, the transmission rates are limited to 64 kbit/s or 1.544 Mbit/s. Even though there is no theoretical reason why frame relay could not operate at, say, 51 Mbit/s, it has not been done. Before this happens, it is likely that newer and faster services such as switched multimegabit data service (SMDS) and broadband ISDN (B-ISDN) will have taken over.

5.4   Switched Multimegabit Data Service (SMDS)

In order to meet the demand for switched data service at multimegabit per second bit rates before the introduction of broadband ISDN, Bell Communications Research, Inc. (Bellcore) designed the *switched multimegabit data service* (SMDS). It is in particular intended to accommodate interconnection traffic for local area networks (LANs).

Figure 5.4 shows an SMDS arrangement. An array of data terminals, personal computers, workstations, printers, and servers is connected to local area networks (LANs). Each LAN has a router that connects the LAN to an SMDS switching system. This switching system is in turn connected to other similar local switching systems and to other such systems operated by interexchange carriers.

The SMDS standard is based on the distributed queue dual bus (DQDB) as described in the IEEE Standard 802.6. The SMDS packet format is described in Sec. 3.6 and shown in Fig. 3.27. Each packet has a fixed length of 53 octets (bytes) with a payload of 44 octets.

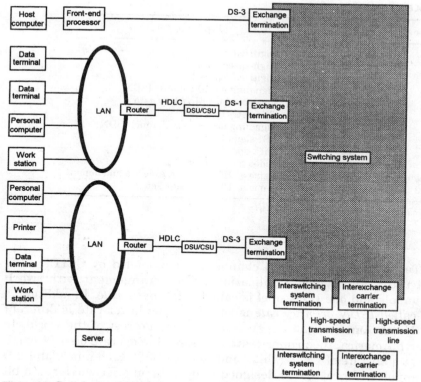

Figure 5.4 Switched multimegabit data service (SMDS) arrangement.

5.5 Integrated Services Digital Network (ISDN)

The main feature of an *integrated services digital network* (ISDN) is that it provides user-to-user digital connections at 64 kbit/s. ISDN also can provide access to packet-switched networks. The initial form of ISDN is now referred to as *narrowband ISDN* and includes channels with bit rates of up to 1.92 Mbit/s. *Broadband ISDN* (B-ISDN) is discussed in the next section.

The idea of ISDN was first conceived by European telecommunications administrations in the mid-1970s. Their thought was to provide for switched digital transmission over existing facilities. In particular, the subscriber loop, i.e., the link between the user and the network, should be on existing twisted copper wire pairs. This limited the available bandwidth and resulted in agreement on the 2B + D arrangement; i.e., two ISDN B channels each carrying a bit stream of 64 kbit/s and an ISDN D channel carrying signaling and some user packet data.

TABLE 5.1   ISDN Channels

Type	Bit rate (kbit/s)	Typical application
B	64	Digitized voice
		High-speed data
		Digitized facsimile
D	16	Signaling for 2B-channel ISDN
		Low-speed user (packet) data
		Telemetry (packet)
D	64	Signaling for 23/30 B-channel ISDN
E	64	SS7 signaling
H0	384	High-speed data
		Video teleconferencing
H11	1536	Same as H0 (in North America and Japan)
H12	1920	Same as H0 (in Europe, etc.)

SOURCE: CCITT Recommendation I.412.

Table 5.1 lists the ISDN channels recommended by CCITT, their bit rates, and typical applications. The arrangement with two B channels and one D channel is called *ISDN basic rate*. In addition, a so-called *ISDN primary rate* is specified. The latter rate is different in North America and Japan from that in the rest of the world. In North America, the primary rate is designed to operate on a T1 carrier at a bit rate of 1.544 Mbit/s and consists of 23 B channels and 1 D channel. In Europe, it is designed to operate on a E1 carrier at a bit rate of 2.048 Mbit/s and consists of 30 or 31 B channels and 1 or 2 D channels.

Six different schemes for coding and multiplexing of the data on the digital subscriber loop (or line) have been presented. A firm worldwide standard is not available. In the United States, the basic access transport system will use two wires with echo cancellation as the mode of transport. The bit rate available for customer use is 144 kbit/s in each direction, consisting of two B channels and one D channel (64 + 64 + 16 kbit/s). In addition, 4 kbit/s will be used to support network operations and 12 kbit/s will be used for framing and timing, for a total of 160 kbit/s in each direction. A four-level code called *2B1Q* (i.e., 2 binary bits contained in a quaternary symbol) is recommended. This is the arrangement that will be used in the United States between the network termination (NT) on the customer premises and the exchange termination (ET) at the network provider. Figure 5.5 shows the basic layout of ISDN and some typical interfaces between user terminals and the network.

At the top, a conventional analog telephone instrument is connected to the integrated services digital network using a conventional

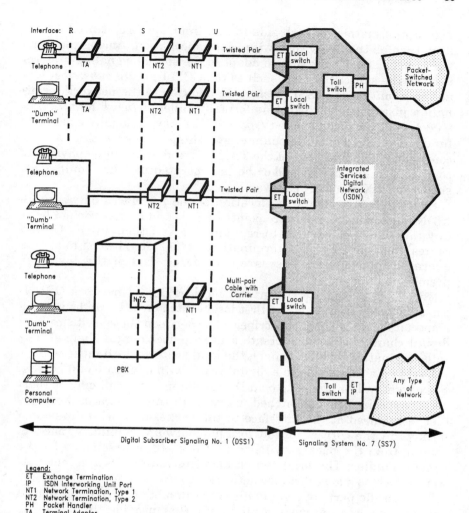

Figure 5.5 Layout of an integrated services digital network.

twisted pair of wires. Because the instrument is analog, a converter is required to digitize the voice and to convert the supervisory and address signals to ISDN standard and vice versa. This device is called a *terminal adapter* (TA).

The digital subscriber loops are terminated by a network termination (NT) at the user's end and by an exchange termination (ET) at the network provider's end of the loop. Their function is to convert the digital information on each side to and from the format used on the loop, such as the 2B1Q code with echo cancellation. Among the functions of NT and ET are also to adapt to the special signaling protocol used on the subscriber loop, such as the CCITT *digital subscriber signaling number one* (DSS1) *system*. In addition to the network termination just discussed, which is formally called *network termination type 1* (NT1), a second called *type 2* (NT2) may optionally exist. The functions of NT2 include channel switching, concentrating, and/or controlling. The functions of the TA, NT1, and NT2 can be physically combined in one device and/or be incorporated in other equipment, such as a PBX or a LAN gateway.

Figure 5.5 also shows the location of the interfaces between the ISDN units, such as reference points R, S, T, and U. They do not have any functions per se. Due to regulatory restrictions in the United States, the type 1 network terminations (NT1) must belong to and be operated by the user. They are considered part of the customer premises equipment (CPE).

Strictly speaking, an integrated services digital network (ISDN) exists between users and the first local switch only. It can be seen as a special form of digital subscriber loop. Between these switches, any digital channel is used to create a user-to-user digital channel. For calls originating on a B channel, the local switch, which must be digital, sets up a connection to a digital trunk with a bit rate of 64 kbit/s. For packet calls originating on a D or B channel, the call may be routed to a special packet-switched network. In this case, a packet handler is required at the interface to the packet network to assemble packets in the correct format for that network. Similarly, packets coming from the packet-switched network are disassembled by the packet handler. The local switch also can route calls to other networks, such as a long-distance network.

The specific parts of signaling system number 7 (SS7) and digital subscriber signaling system number 1 that provide the signaling means for specific services are the user parts. Early implementations of SS7 had a telephone user part (TUP), and later versions had an integrated services digital network user part (ISDN-UP). The ISDN-UP is prevalent in North America, while the rest of the world still uses the TUP. Both user parts can coexist in a network. Basically, ISDN-UP is an enhancement that provides services for nonvoice calls and other ISDN services, as well as intelligent network (IN) services. The types of services increase steadily. Through enhancement of the computing power in the SS7 network, many advanced services can be added, including network-wide call forwarding, flexible allocation of bandwidth, etc. When all this is included, we have an advanced intelligent network (AIN).

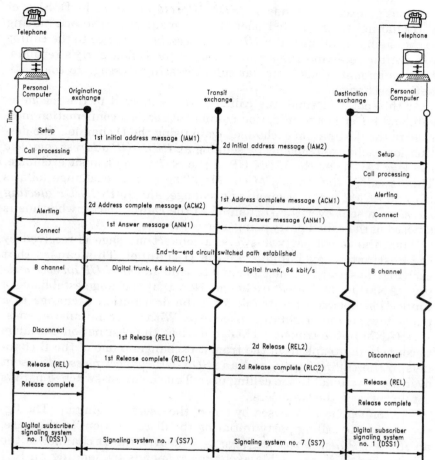

Figure 5.6 DSS1 and SS7 ISDN-UP call setup example.

Among the basic services provided by ISDN-UP is the control of user-to-user circuit-switched connections. Figure 5.6 shows the procedure and the ISDN-UP messages used to set up a typical connection between two users through an originating local exchange, a transit exchange, and a destination local exchange. The calling user activates a setup message to the originating exchange using the *ISDN user part* (ISDN-UP) of DSS1 on the D channel. That exchange sends an *initial address message* (IAM) to the transit exchange, the first such message (IAM1), using the ISDN-UP of SS7. The transit exchange sends a second IAM (IAM2) to the destination exchange, still using the ISDN-UP. On finding the called party free to accept a call, the destination exchange sends an *ISDN-UP setup message* by DSS1 over a D channel to the called user. In the meantime, the originating exchange has sent an *ISDN-UP call-processing message* to the calling user. After receiving the setup message, the called party's termination equipment sends a similar call-processing message to the destination exchange.

At this point in time, the called party is alerted to the incoming call, and his or her termination equipment sends a confirmation message to the destination exchange, again using the D channel. The destination exchange responds by sending an *ISDN-UP address complete message* (ACM) by the ISDN-UP of the SS7 to the transit exchange. This is called the *first ACM* (ACM1). The transit exchange follows with an ACM2 to the originating exchange, and an *ISDN-UP alerting message* is sent over the D channel to the calling party, who thus is advised of the called party being alerted.

When the called party answers, a connect message is received by the destination exchange over the D channel. This causes that exchange to set up a connection and to send an *ISDN-UP answer message* (ANM) to the transit exchange. Now that exchange establishes a connection between the trunk from the destination exchange to a trunk toward the originating exchange. With a second answer message (ANM2), the transit exchange advises the destination exchange to connect the trunk from the transit exchange to one of the B channels to the calling party, and an *ISDN-UP connect message* is sent over the D channel to the calling user. Thus a circuit-switched path is set up between the two users.

The connection is released by one of the users hanging up. The figure shows the calling party initiating the disconnection. In this case, an *ISDN-UP disconnect message* is sent over the D channel to the originating exchange. This exchange responds by sending an SS7 *ISDN-UP release message* (REL) toward the transit exchange, which continues with another REL toward the destination exchange. Then an ISDN-UP disconnect message is sent to the called party's equipment by DSS1 over the D channel. After having released their parts of the connection, the transit exchange and the destination exchange send SS7 *ISDN-UP release complete messages* (RLC) toward the calling party. The calling user's equipment sends an ISDN-UP RLC over the D channel to the originating exchange, and the destination exchange sends a similar message to the called user's termination equipment. After this, all connections are released and available for other calls.

Among special services are the so-called Freephone, which uses the 800-number series in North America. In this case, a special data base has to be interrogated before a call can be set up. This data base has a list of 800 numbers (or their equivalent in other countries) with the corresponding physical numbers. The physical numbers can differ with the origination of the call or with the time of day. Thus calls can be directed to the closest serving facility or to a facility in a different time zone in order to facilitate off-hour calls. The data bases for Freephone numbers are associated with network control points. Figure 5.7 shows how such calls are set up. The major difference

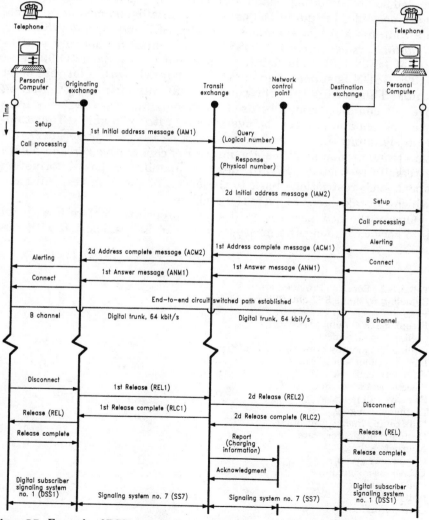

Figure 5.7   Example of DSS1 and SS7 setup of Freephone (800-number) calls.

between this figure and Fig. 5.6 is that the call setup procedure is interrupted by the interrogation of the data base at a network control point. In the case in the figure, the transit exchange sends an *ISDN-UP query message* with the logical number to the network control point, and the physical number is sent back with an *ISDN-UP response message*. At the end of the call, the transit exchange reports the charging information to the network control point, which acknowledges receipt of the information.

Besides the basic call setup service and the Freephone service discussed so far, signaling system number 7 (SS7) makes many new services possible, including the creation of intelligent networks. Table 5.2 contains a list of examples. Further, SS7 and *digital subscriber signaling system number 1* (DSS1) are essential for network switching of ISDN calls. The DSS1 is specified in standards Q.920 and Q.921. ISDN is supposed to offer clear channels with a bit rate of 64 kbit/s. This means that the conventional way of in-band signaling over T1 channels cannot be used. Such signaling "borrows" bits from the channel and reduces the available bit rate to 56 kbit/s. Thus common signaling outside the channels is a necessity. Further, over certain transmission facilities, the number of consecutive zeros in a data stream is restricted. SS7 provides the capability to pass information about such restrictions back to the origin of the data transmission, which can take steps to avoid the problem.

One of the new services created an uproar even before it was put into service. Among the messages that can be transported with sig-

TABLE 5.2   Services Provided for by Signaling Systems SS7 and DSS1

Freephone (800-number service)
Calling line identification
Calling line identity presentation (CLIP)
Blocking caller identification
Automatic callback
Selected call forwarding
Selective call rejection
Distinctive ringing
Bulk calling line identification
Customer originated trace
Alternate billing
Closed-user group
User-to-user signaling
Call forwarding
Call hold
Call waiting
Conference calling

naling system number 7 (SS7) is the physical number of the calling party, called *calling line identification*. This information can be displayed to the called party, a service called *calling line identity presentation* (CLIP). Through look-up in a data base, not only the calling number but the name listed in the directory can be displayed. The benefits to a user is that he or she will know who is calling before accepting the call. On calls to companies, the agent answering the call can have information about the caller displayed on a screen when the call is answered. People in telemarketing think that it is a great idea that they can answer the call with "Good afternoon Mr. Brown." However, what happens if it is Mr. Brown's wife or child that calls? People in the insurance industry say that the agent can have the caller's portfolio in front of him or her when starting the conversation. The same goes for the financial and travel industries. On the other hand, some people say that this is an invasion of privacy and are trying to stop the implementation of this service. For example, people with unlisted telephone numbers do not want to have them displayed. In response, the telephone industry states that display of the caller's identity can be prevented by technical means. A customer who does not want his or her telephone number displayed can have the data base instructed to that effect.

Once calling line identification is in effect, it can be used for many purposes. One is *automatic callback*; i.e., the calling party is automatically called when both lines are free. Another is *selective call forwarding*; i.e., calls are forwarded depending on the identity of the caller. If a user is away from his or her home or office, he or she may want calls forwarded to another physical number. Depending on how far away from the location of the directory number the user is, he or she may not want to incur the charges for all forwarded calls. In this situation, selective call forwarding enables the customer to select the calls to be forwarded. Calls from a lover are desired but not those from the ex-mother-in-law.

Calling line identification also can be used for rejection of selected calls. With a "black list" built into the system, calls from numbers on that list can be rejected by the system. This service is called *selective call rejection*.

Similarly, the system can be programmed to alert the called party differently depending on who is calling. Different kinds of ringing, buzzing, etc., can be used, as well as distinctive voice messages. Rather than having the telephone ring, we might hear it say, "John is calling." The name for this service is *distinctive ringing*.

A version of calling line identification allows users of Centrex, multiline hunt groups, and private branch exchanges (PBXs) to receive lists of calls from outside these services indicating the physical number that called, the time of the call, etc. The name for this service is *bulk calling line identification.*

Signaling system number 7 also allows for the tracing of nuisance calls. The system will pick up the number of the calling and called persons and the time and duration of the call. With this service, *customer-originated trace,* the user of the service and/or a law enforcement agency will get reports on the calls.

Because different data bases can be interrogated through signaling system number 7, it is possible to verify and authorize *alternate billing* of calls. Thus the validity of a calling or credit card number can be checked. Billing to a third party and collect calls also can be automated.

Since the system can keep track of the origin and destination of every single call, it can be programmed so that specific lines only can be connected. In other words, a private network is obtained or a *closed user group* created.

The signaling capabilities of DSS1 and SS7 can accommodate *user-to-user signaling.* For example, user information can be incorporated in the signaling messages transmitted over the signaling network.

Services such as *call waiting, call forwarding,* and putting *calls on hold* are available in different forms. Many telephone sets have "hold" buttons. One drawback with this arrangement is that the line is kept occupied. Several providers of telephone service offer features that alert you about a waiting call and let you switch back and forth between two calls. Telephone companies also offer forwarding of calls. By using the signaling capabilities of the digital subscriber signaling system (DSS1), these features can be handled by the local switch without keeping the subscriber lines occupied.

The setting up of *conference calls* is also made more efficient through use of the new signaling systems. Many other services are possible. Bellcore, the research organization of the U.S. regional Bell operating companies, is planning an *advanced intelligent network* (AIN). The software of new digital switches allows for interruptions in call setups, during which data bases can be interrogated via SS7 to modify the call setup and accommodate new features. Such a switch is called a *service switching point* (SSP). Depending on the requirements of the service being provided, call processing by an SSP switch may continue (in which case the message is called an *event report*) or be suspended (in which case the message is called a *query*) to await return instructions from a signaling or network control point. Event reports can be used to coordinate off-network events, such as data communication events, with incoming or outgoing calls. They also can provide distinctive ringing. Responses to queries can be used to transfer calls, for example. There do not seem to be any limits on what may be proposed and implemented. The future will see many more services than those discussed here.

5.6 Broadband Integrated Services Digital Network (B-ISDN)

The integrated services digital network (ISDN) concept is being broadened to allow for bit rates above 2 Mbit/s. The new concept, broadband integrated services digital network (B-ISDN), is being formalized within international and national standards bodies. International standards are being formed by the Telecommunications Standardization Committee of the International Telecommunication Union (ITU), a United Nations agency. Until recently, the ITU-T was known as the CCITT (*Comité Consultatif International Télégraphique et Téléphonique,* or International Telegraph and Telephone Consultative Committee). In 1990, CCITT approved 13 recommendations regarding B-ISDN. The relevant ones are listed in the References section of this chapter. Bell Communications Research, Inc. (Bellcore), the joint research company of the U.S. Bell operating companies, has issued advisories regarding B-ISDN. The relevant ones are also listed in the References section of this chapter.

A variety of terminals can be connected to a broadband ISDN network. Figure 5.8 shows conventional telephones, digital telephones, personal computers, workstations, facsimile machines, and printers. One of the purposes of B-ISDN is to provide video-on-demand to television-type sets. Terminals can be connected to the B-ISDN network directly or through other networks, such as local area networks (LANs).

As is the case with the narrowband cousin, the line from the broadband network ends with a broadband network termination type 1 (B-NT1). Besides terminating the line transmission, the functions of the B-NT1 include transmission interface handling and operations, administration, and maintenance (OAM) functions.

On the "other side" of the B-NT1 we find a broadband network termination type 2 (B-NT2). Its functions are similar to those of the narrowband NT2 unit, i.e., switching and distribution of internal connections (such as a PBX), adaptation to different media and topologies, resource allocation, concentration, buffering, multiplexing, and demultiplexing. A B-NT2 also has signaling, operations, administration, and maintenance (OAM) and interface functions. In order to interface with terminal equipment that is not designed for ISDN or B-ISDN, special terminal adapters (TA and B-TA) are required.

Broadband network termination types 1 and 2 have little more than part of the name in common with the conventional narrowband

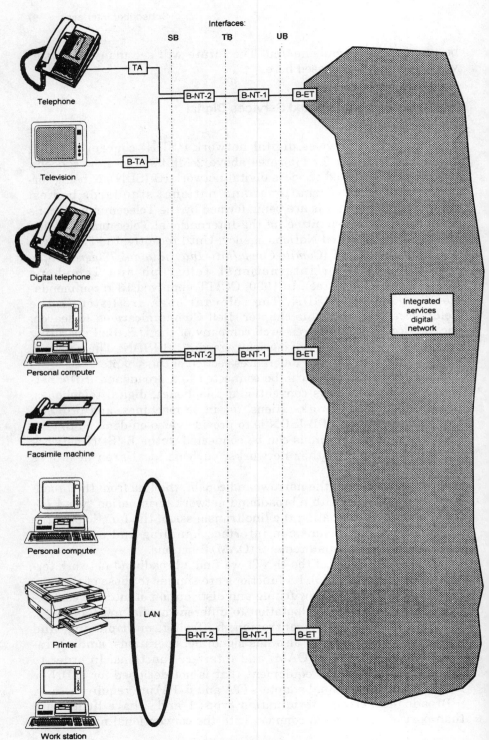

Figure 5.8 Different terminals interfacing with a broadband integrated services digital network.

NT1s and NT2s. While the latter interface with a circuit-switched network, the B-NT1s interface with a packet-switched network or rather a network operating in the asynchronous transfer mode (ATM). When transmitted between the TE and the NT1 in the case of narrowband ISDN, data belonging to the two B channels and the D channel are interwoven in the form

$$B1 \ D \ B2 \ D \ B1 \ D \ B2 \ D \ B1 \ D \ B2...$$

where each B-channel group consists of 8 bits and each D-channel group consists of 1 bit. Over the local loop, i.e., between the NT1 and the central office and across the U reference point, the signaling scheme used is called 2B1Q for "two binary, one quaternary." This is a four-level line code in which a pair of bits defines a single quaternary symbol (quat).

According to Bellcore's technical advisory regarding broadband-ISDN user-to-network interfaces (TA-NWT-00112), these interfaces are established at 1.544, 44.736, 51.840, 155.520, and 622.080 Mbit/s. The first two, at 1.544 and 44.736 Mbit/s, correspond to conventional digital signal levels one and three (DS-1 and DS-3). They are included in the advisory so that B-ISDN services can be accessed with existing DS-1 and DS-3 digital facilities. The three highest bit rates correspond to the SONET STS-1 at 51.840 Mbit/s, the SONET STS-3c at 155.520 Mbit/s, and the SONET STS-12c at 622.080 Mbit/s. SONET is an acronym for *synchronous optical network,* and STS stands for *synchronous transport signal.* They are discussed further in Chap. 8.

The current Bellcore advice is that the user-network interface in B-ISDN shall consist of two single-mode optical fibers, each transmitting point-to-point in one direction. The use of wavelength-division multiplexing (WDM) in order to accomplish transmission in both directions on the same fiber is under consideration by Bellcore and other organizations. The feasibility of using asymmetric transportation, for example, 622.080 Mbit/s in one direction and 155.520 Mbit/s in the other, is being studied. It is not so much a technical problem as a question of whether the market will require asymmetric transports.

The ITU Telecommunications Standardization Sector (ITU-T) is working on a draft of an international standard for B-ISDN layer 3 signaling referred to as draft ITU-T Recommendation Q.2931 (formerly Q.93B). It specifies the layer 3 call/connection states, messages, information elements, timers, and procedures used to control B-ISDN connections. These procedures apply at the interface between B-ISDN terminal equipment (including a private network) and a B-ISDN public network (at reference points S_B and T_B). Table 5.3 lists the basic

TABLE 5.3   Basic Capabilities Supported by ITUT Recommendation Q.2931 (Q.93B)

1. Demand (switched) channel connections
2. Point-to-point switched-channel connections
3. Connections with symmetric or asymmetric bandwidth requirements
4. Single-connection (point-to-point) calls
5. Basic signaling functions via protocol messages, information elements, and procedures
6. Class X, class A ATM transport services
7. Request and indication of signaling parameters
8. Negotiation of virtual path connection identifier (VPCI), virtual path identifier (VPI), and virtual channel identifier (VCI)
9. A single, statistically defined out-of-band channel for all signaling messages
10. Error recovery
11. Public UNI addressing formats for unique identification of ATM endpoints
12. End-to-end compatibility parameter identification
13. Signaling interworking with N-ISDN and provision of N-ISDN services
14. Support of supplementary service

capabilities supported by the ITU-T draft Recommendation Q.2931 (formerly Q.93B).

As listed in Table 5.3, a *point-to-point switched-channel connection* refers to a collection of virtual channel (VC) or virtual path (VP) links that connect two endpoints. Connections can be asymmetric in the sense that the bandwidth specified in the two directions differs. Class X refers to a connection-oriented ATM transport service where the ALL, traffic type [variable bit rate (VBR) or constant bit rate (CBR)], and timing requirements are user defined. For each connection, the following compatibility parameters can be specified: type of ALL (1, 3/4, or 5), method of protocol multiplexing (LLC versus VC), and ALL parameters and protocols above the network layer.

Table 5.4 lists the types of signaling messages available for transmission between users and the network. They are transmitted in binary form (octets), usually on a separate, signaling virtual channel (SVC), i.e., not via the common signaling network (e.g., SS7).

Table 5.5 lists and describes the signaling messages sent between users and the network for B-ISDN call and connection control. Similar but different messages are used in support of 64 kbit/s based ISDN circuit mode services. In addition, *restart* and *restart acknowledge* messages are used to restart indicated virtual channels and paths.

The information elements containing the ATM adaptation layer (AAL) parameters are listed in Table 5.6.

User-to-user signaling as a supplementary service to B-ISDN is specified in ITU-T draft Recommendation BQ.957.1. Digital subscriber signaling system number 2 (DSS2) is the means for this ser-

TABLE 5.4   Types of Signaling Messages

Message	Description
Protocol discriminator	Distinguishes user-network call control messages from others
Call reference	Identifies the call to which a message applies
Message type	Identifies the function of the message being sent; the types of messages are listed in Table 5.5
Message length	Identifies the length of the contents of a message

Information elements	
ATM adaptation layer (AAL) parameters	Contains the AAL parameters requested by a user for a call; Table 5.6 lists the types
ATM user traffic description	Indicates the peak cell rate for user information
Broadband bearer capability	Indicates the broadband bearer capability requested from the network by the user, such as bearer class, traffic type (CBR versus VBR), timing requirements, susceptibility to clipping, and user plane connection configuration (e.g., point-to-point)
Broadband high-layer information	Used to check the type of high-layer protocol to use
Broadband low-layer information	Used to check the type of low-layer (layers 1, 2, and 3) protocol to use
Call state	Describes the current state of a call/connection
Called party number	Identifies the called party's number
Called party subaddress	Identifies a specific terminal, etc., of the called party
Calling party number	Identifies the calling party's number
Calling party subaddress	Identifies a specific terminal, etc., of the calling party
Connection identifier	Identifies the virtual path and channel in a connection
End-to-end transit delay	Requests and indicates the nominal maximum permissible delay in a call
Quality of service (QOS) parameter	Indicates a certain QOS class
Broadband repeat indicator	Indicates how to interpret repeated information elements
Restart indicator	Identifies the class of the facility to be restarted
Broadband sending complete	Indicates the completion of called party number
Transit network selection	Identifies a requested transit network
Notification indicator	Indicates information pertaining to a call
OAM traffic descriptor	Describes the operations, administration, and maintenance (OAM) traffic

TABLE 5.5   Messages for B-ISDN Call and Connection Control

Message	Description
Alerting	Called user alerting has been initiated
Call proceeding	Indicates that the requested call establishment has been initiated
Connect	Indicates call acceptance by the called user
Connect acknowledge	Sent by the network to the called user to indicate that the connection is acknowledged
Release	Sent by the user to request the network to release the connection; also sent by the network to indicate that the connection is cleared
Release complete	Indicates that the release of a connection is completed
Setup	Sent by the calling user to the network and by the network to the called user to initiate B-ISDN call and connection establishment; this message contains information about the type of call, etc.
Status	Sent by the user to the network in response to a "status inquiry" message or to report certain error conditions
Status inquiry	Sent by the user or the network to solicit a "status" message
Notify	Sent by the user or the network to indicate information pertaining to a call/connection

TABLE 5.6   ALL Parameters Information Elements

Length of ALL parameter contents

ALL type

ALL subtype

Constant bit rate (CBR)

Multiplier

Source clock frequency recovery method

Error-correction method

Structures data transfer block size

Partially filled cells method

Forward maximum common part convergence sublayer-service data unit (CPCS-SDU) size

Backward maximum common part convergence sublayer-service data unit (CPCS-SDU) size

Message identifier (MID) range (lowest MID value)

Message identifier (MID) range (highest MID value)

Service specific convergence sublayer (SSCS) type

User-defined AAL information

AAL type

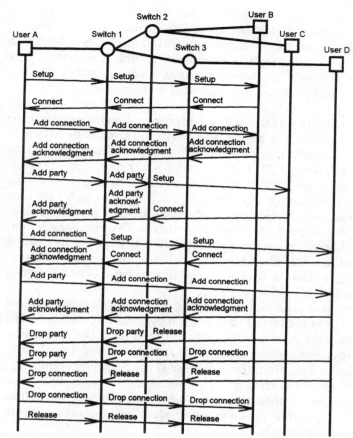

Figure 5.9 B-ISDN multiparty multiconnection setup.

vice. This is the broadband version of the DSS1 for N-ISDN. DSS2 allows a B-ISDN user to send and/or receive a limited amount of information to and/or from another B-ISDN user over the signaling virtual channel in association with a call/connection to the other B-ISDN user.

Figure 5.9 shows an example of a multiparty multiconnection setup. User A sets up a bidirectional point-to-point connection with user B by way of switches 1 and 3. *Setup* messages (see Table 5.5) are sent from user A to switches 1 and 3, and *connect* messages are returned. User A then requests a second connection, a point-to-multipoint connection, to user B, also by way of switches 1 and 3, sending *add connection* messages, and *add connection acknowledgment* messages are returned.

User A adds user C to the point-to-multipoint connection. An *add party* message is used as the connection from user A to switches 1 and 2 exists. Since no common path is available between switch 2 and user C, a *setup* message is sent from switch 2 to user C. In response, user C sends a *connect* message back to switch 2. This switch sends an *add party acknowledgment* message back to switch 1 and user A.

Now user A adds user D through a bidirectional point-to-point connection. An *add connection* message is sent from user A to switch 1. This is converted to a *setup* message, which is sent from switch 1 to switch 3 and on to user D. This is acknowledged through user D sending a *connect* message back to switches 3 and 1. Switch 1 sends an *add connection acknowledgment* message back to user A. A new connection between users A and D is established. User A then adds user D by sending an *add party* message to switch 1. This message is converted to an *add connection* message by switch 1, which is forwarded to switch 3 and user D, who responds with *add connection acknowledgment* messages back to user A. A call is now established between user A and users B, C, and D.

User C drops from the call by sending a *release* message toward switch 2, where it is converted to a *drop party* message, which is sent to switch 1 and user A. Similarly, user D drops from the connection by sending a *drop connection* message to switches 3 and 1 and on to user A. In order to drop from the call, user D sends a *release* message to switch 3 and on to switch 1. Switch 1 sends a *drop connection* message to user A. User A then clears the connection to user B by first sending *drop connection* messages and then *release* messages to switches 1 and 3 and user B.

5.7   Commercial Asynchronous Transfer Mode (ATM) Gateways

Asynchronous transfer mode (ATM) will be the basis for broadband networks of the future. Among functions that must be handled at the interface between users and ATM networks are concentration of multiple services into one broadband link, cell reassembly and segmentation (assembly and disassembly), conversion between electrical and optical circuits, and establishment of the desired type of connection.

The SMX-6000 ATM service multiplexor from Fujitsu Network Switching of America (FNS) shown in Fig. 5.10 is designed as an ATM gateway. It provides access to public and private ATM networks at bit rates of 1.5, 45, and 155 Mbit/s over electrical or optical links. Several different types of data streams can be multiplexed onto the same 155-Mbit/s channel. Through the use of different interface

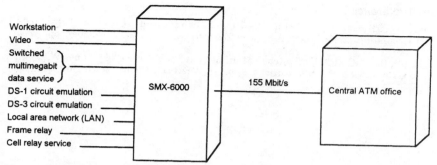

Figure 5.10 Fujitsu SMX-6000 ATM service multiplexor.

cards, an SMX-6000 can be configured to handle DS-1 and DS-3 circuit emulation, switched multimegabit data service (SMDS), frame-relay service and cell-relay service. The SMX-6000 can be used as an add and drop multiplexor, as described in Sec. 8.3.

References

Standards

Bellcore Technical Reference TR-TSV-000772, issue 1 (1991), "Generic system requirements in support of switched multimegabit data service," Bellcore, Livingston, N.J.

Bellcore Technical Reference TR-TSV-000773, issue 1 (1991), "Local access system generic requirements, objectives, and interfaces in support of switched multimegabit data service," Bellcore, Livingston, N.J.

CCITT Recommendation I.121 (1991), "Broadband aspects of ISDN," CCITT, Geneva.

CCITT Recommendation I.150 (1991), "B-ISDN asynchronous transfer mode functional characteristics," CCITT, Geneva.

CCITT Recommendation I.361 (1991), "B-ISDN ATM layer specification," CCITT, Geneva.

CCITT Recommendation I.362 (1991), "B-ISDN ATM adaptation layer (AAL) functional description," CCITT, Geneva.

CCITT Recommendation I.413 (1991), "B-ISDN user-network interface," CCITT, Geneva.

Bellcore Framework Technical Advisory FA-NWT-001109, issue 1 (1990), "Broadband ISDN transport network elements framework generic criteria," Bellcore, Livingston, N.J.

Bellcore Framework Technical Advisory FA-NWT-001110, issue 1 (1990), "Broadband ISDN switching system framework generic criteria," Bellcore, Livingston, N.J.

Bellcore Technical Advisory TA-NWT-001112, issue 1 (1992), "Broadband-ISDN user to network interface and network node interface physical layer generic criteria," Bellcore, Livingston, N.J.

Bellcore Technical Advisory TA-NWT-001113, issue 2 (1993), "Asynchronous transfer mode (ATM) and ATM adaptation layer (AAL) protocols generic requirements," Bellcore, Livingston, N.J.

ITU Telecommunications Standardization Sector, Draft Recommendation Q.93B (1993), "Broadband ISDN user-network interface layer 3 specification for basic call/connection control," ITU, Geneva.

ITU Telecommunications Standardization Sector, Draft Recommendation BQ.957.1 (1993), "User-to-user signaling (UUS)," ITU, Geneva.

Other references

Lindberg, Bertil C. (1991), "Switched Broadband Network Interfaces," *Telecommunications* (North American Edition), vol. 25, no. 10, October, pp. 47–51, 53, 58.

Lindberg, Bertil C. (1993), "Switched Multimegabit Data Service," in James W. Conard (ed.), *Broadband Communications Systems,* Auerbach Publications, New York, pp. 163–173.

6

Network Management Technology

6.1 Introduction

From the old situation where the service providers controlled the networks exclusively, the trend is toward more and more user management. In dial-up services, the users always controlled the dialing of the address of the called party. Users also controlled the initiation and clearing of connections. Theses features remain today and will be there in the future.

Users also have the opportunity to select the type of network, such as telephone, telex, packet switching, or facsimile, for a connection. The future trend is toward one single network to handle all these different services, the B-ISDN. Still, users will have to select bandwidth, quality of service, routing, etc.

6.2 Addressing

Conventional telephone calls are addressed by a telephone number consisting of access code, country code, area code, exchange prefix, and an individual user number. Within a PABX area, only the individual number has to be dialed. The further away from the local switch area one is, the more numbers have to be added. Figure 6.1 shows an example of an international telephone number. The number starts with 011, which is the access code to international calls from the United States. It is followed by a country code, in this case the United Kingdom, and an area code, in this case Birmingham. A *prefix* typically indicates an exchange. Note that there may be more than

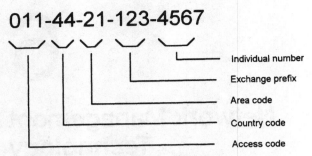

Figure 6.1 International telephone number.

one exchange in a central office. Each of the groups—country code, area code, prefix, and individual number—can range in length from one to several digits.

The international numbering plan for integrated service digital networks (ISDNs) is similar to the one for conventional voice traffic. It is described in CCITT (ITU-T) Recommendations E.164 (= I.331), I.330, I.331, I.332, I.333, and I.334.

A slightly different terminology is used for ISDNs, where an ISDN *number* indicates the network number and an ISDN *address* adds some digits that identify a specific terminal. Packet-switched networks use a different numbering plan to identify hosts on the network. This numbering plan is shown in Fig. 6.2. This plan is specified in CCITT (ITU-T) Recommendation X.121.

An E-mail address consists of a user name, a computer base, an organization, and a type of organization. Each is called a *domain* in the U.S. *Internet* system (Fig. 6.3). The highest-level domain, i.e., the last group, indicates the type of organization housing the E-mail address in the United States. Table 6.1 shows the original Internet high-level domains. For international E-mail, the highest-level domain indicates the country. Thus CA indicates Canada; FR,

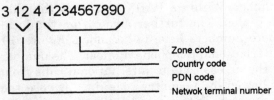

Figure 6.2 Numbering plan for packet-switched public data networks (PSPDNs).

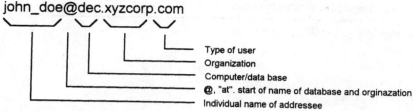

Figure 6.3 E-mail address.

France; DE, Germany; MX, Mexico; UK, United Kingdom; and US, the United States.

Some networks have their own names either at the highest level or next to the highest level. Among those using their name at the highest level are *bitnet* and *UUCP*. Using the second-highest domain level are *CompuServe, fodonet, sprint,* and *mcimail,* followed by the *com* domain (such as @compuserve.com).

All Internet addresses and most other ones have a corresponding digital number that is used internally by the networks to find an addressee. Such a number may look like 192.112.36.5 or 128.174.5.6.

6.3 Broadband Services

In the areas of broadband services, broadband integrated services digital network (B-ISDN) and asynchronous transfer mode (ATM) users can negotiate with the network and establish the types of services required by the user and offered by the network. Table 6.2 lists such negotiable services. Some classes of service are prearranged and listed in Table 6.3. The signaling media available for these service negotiations are discussed in Sec. 5.6 above.

TABLE 6.1   U.S. Internet High-Level Domains

Domain	Usage
com	Businesses and other commercial organizations
edu	Universities and other educational organizations
gov	Nonmilitary government organizations
mil	Military organizations
org	Other, mainly nonprofit organizations
net	Network resources

TABLE 6.2   Negotiable B-ISDN Services

ATM adaptation layer (AAL) parameters

Bandwidth

Error handling

High-layer protocols

Low-layer protocols

Maximum transmission delays

Peak cell rate

Quality of service

Routing (requested transit network)

Susceptibility to clipping

Timing requirements

Traffic type [constant bit rate (CBR) versus variable bit rate (VBR)]

Type of connection (connection-oriented versus connectionless)

User plane connection configuration (e.g., point-to-point, point-to-multipoint, broadcast)

6.4   User-Network Signaling

Signaling is used to control communications networks, i.e., to establish, maintain, and release connections. It is also used for billing purposes and to make certain services available to a user. Conventional signaling between users and central offices consists of supervisory signals, i.e., off-hook/on-hook signals, and numerical address (dialing) signals. When these terms were coined, the telephone really had a hook from which the receiver was lifted when you wanted to make a call and on which the receiver was replaced after the end of the call. Similarly, telephones had rotary dials, and some still have.

New services will require additional types of network signaling. A user may not just want to send the numerical address of the called party to the network; he or she also may want to know the numerical address of the calling party. Users also may want to convey information to the network about how the call should be routed. Should it

TABLE 6.3   Classes of Broadband Services

Class	Timing	Bit rate	Connection mode
A	Yes	Constant	Connection-oriented
B	Yes	Variable	Connection-oriented
C	No	Variable	Connection-oriented
D	No	Variable	Connectionless
X	User-defined	User-defined	Connection-oriented

be handled by AT&T, MCI, Sprint, or some other long-distance carrier, and should it be carried on a voice link or a digital switched-data link or be packet switched, for example? Users also may want to send signals between them. We have all heard things like "If you have a touch-tone phone and want to talk to a live person, press '0,' now."

Originally, signals from the user to the local telephone switch were sent in the form of interruptions of the direct current in the subscriber loop. Touch-tone signals consist of two tones with selected, distinct frequencies that are sent simultaneously. Twelve different combinations are standard, the 10 digits plus star (*) and pound (#). Some, mainly military, instruments have 16 combinations. For modern digital systems, such as for the different services available on an integrated services digital network (ISDN), 12 or even 16 different signals are not enough. Here signals are sent in the form of binary codes with up to 256 bytes (2048 bits), which allows for that many pieces of information.

Similar codes are used in data links between computers and their terminals. These types of signals are called *protocols*. Basically, signaling is conducted according to a specific protocol. Among such protocols are the link-access procedure (LAP), the versions of balanced link-access protocol (LAPB) used in X.25, and the link-access procedure/protocol on the D channel (LAPD) used for ISDN. X.25 is a recommendation by the International Telegraph and Telephone Consultative Committee (CCITT), now ITU-T. Recommendations of the CCITT have the status of international standards for international telegraph and telephone. CCITT is part of the International Telecommunications Union (ITU), which, in turn, is an entity of the United Nations. CCITT Recommendation X.25 specifies the "interface between data terminal equipment (DTE) and data-circuit terminating equipment (DCE) for terminals operating in the packet mode and connected to public data networks by dedicated circuit[s]." CCITT Recommendation X.75 covers the interface or gateways between separate packet-switched networks.

Signaling data according to CCITT Recommendations X.25 and X.75 are sent in the form of packets. Each packet (in most cases synonymous with a frame) has a flag, an address, a control sequence, an information sequence, and another flag. The flags are there for synchronization and to determine the beginning and end of a frame. Typically, they consist of 1 byte (equal to 8 bits). The address indicates where the packet shall be sent. The control sequence is there for controlling information transfers and for supervisory functions. In addition, each packet can contain information. These sequences consist of varying numbers of bytes. Figure 6.4 shows a typical frame or packet.

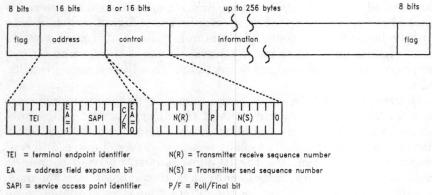

Figure 6.4 Typical data frame/packet.

6.5 Internal Network (Node-to-Node) Signaling

Besides between users and the network, signaling is also performed between elements of the network—in particular, between switching machines. Contained in this signaling is information on routing, available links, whether the called party is available or not, whether he or she is being alerted, etc. Such signals can be transmitted within the (voice) channel, in the edges of the channel, or on completely separate channels. These forms are called *in-band signaling, out-of-band signaling,* and *common-channel signaling* (CCS), respectively. Common-channel signaling is conducted on a separate packet-switched network.

The American Telephone and Telegraph Company (AT&T) introduced the North American common-channel interoffice signaling (CCIS) system in 1976. It is similar to and partly compatible with the CCITT standard signaling system number 6 (SS6).

As mentioned in Chap. 1, CCITT has introduced a new common-channel signaling (CCS) system. It is specified in CCITT Recommendations Q.700-716, Q.721-766, and Q.771-795. Upgrading of the North American common-channel signaling system to a version of CCITT SS7 is in progress.

All data communications protocols, including the signaling protocol of SS7, can be fitted into the open systems interconnection (OSI) standards issued by the International Organization for Standards (ISO), headquartered in Geneva. These standards are covered in CCITT Recommendations X.200 through X.250. The OSI recommendation organizes data communications protocols in *layers* numbered from 1 to 7, as shown to the left in Fig. 6.5. Layer 1, the OSI *physical layer,*

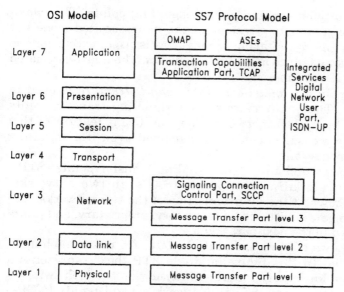

Figure 6.5 The OSI model and the SS7 protocol model.

describes the electrical, mechanical, functional, and procedural characteristics of an interconnection. Layer 2, the OSI *data link layer,* is responsible for the transport and error-free delivery of data over the physical link. The OSI *network layer,* layer 3, is responsible for the transparent transport of data and provides a means for establishing, maintaining, and releasing connections. This layer controls switching and selects the proper route. Layers 1 through 3 provide the network for transporting messages. Layer 4, the OSI *transport layer,* is responsible for the transport of data which are supplied by the higher layers over the facilities of the lower layers. The OSI *session layer,* layer 5, establishes "sessions" between the users. Layer 6, the OSI *presentation layer,* defines the method of data assembly and ensures that the data are presented to the applications layer in an acceptable form. The OSI *applications layer,* layer 7, supports the users' application programs. We will return to the right part of Fig. 6.5.

In the signaling networks of common-channel signaling system number 7 (SS7), signals are sent between signaling points over certain links. Three different signaling modes exist and are defined by the links between the signaling points. In the *associated mode,* the signaling messages are sent over a link that directly interconnects two signaling points. In the *nonassociated mode* of signaling, messages are sent over tandem links that pass through signaling points other than the originating and ending signaling points. The *quasi-associated mode* of signaling resembles the nonassociated mode but for the fact that the route taken by messages is fixed, except for the rerouting caused by failure and recovery events. The quasi-associated mode is used in North America.

More specifically, the *quasi-associated quad structure* is used in North America. The *quad* refers to four signaling transfer points interconnected by a mesh network. The significant feature of this structure is the messages can be transported even if a signaling transfer point or link fails.

Figure 6.6 shows a chain of six signaling transfer points (STPs) connecting four signaling end points (SEPs) in two networks. Gateways are shown at the interconnection of the two networks. The reasons for such gateways (such as regulatory, proprietary, and protocol) are discussed in the next paragraph.

Figure 6.7 shows a more elaborate layout of a common-channel signaling system. In this figure, a single STP "box" represents a *quad*. The dotted lines carry signaling information between switches via signal transfer points (STPs). Network control points (NCPs) support a wide range of network applications. They are discussed below. Associated with the NCPs are different types of data bases. These contain information about the routing of calls to specific subscribers, e.g., in the case of 800 numbers in the United States (Freephone in other parts of the world), calls to be forwarded, authorization of credit-card billing, etc. In the United States, the operating areas of individual providers of telecommunications services are limited by regulatory constraints. The freedom of one provider to directly inquire from a data base in another provider's area is limit-

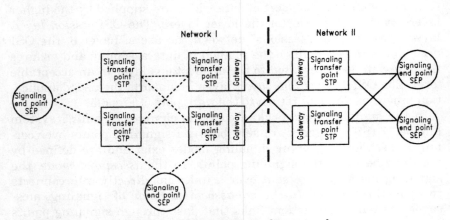

Figure 6.6 Interconnection of two SS7 common signaling networks.

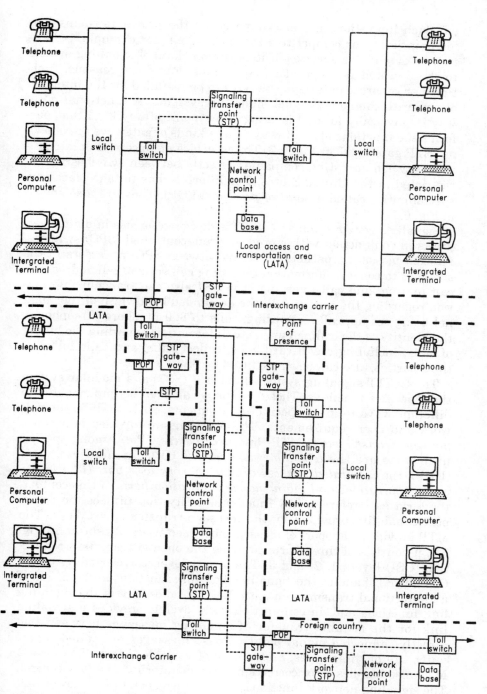

Figure 6.7 Common-channel signaling network.

ed. Such limitations are also imposed by the providers themselves for proprietary or competitive reasons. AT&T, for example, is hesitant to extend the services of its common-channel signaling system to the divested Bell operating companies. For these reasons, functional or separate gateways will have to be installed for the interface of separate common-channel signaling systems. A functional gateway is physically located in a signaling point of any kind that performs the functions of a gateway. These kinds of gateways are shown as "STP gateways" in Fig. 6.7. The interface of voice and data channels between operating companies directly between switches is also restricted in the United States. The interface has to take place at a special entity called a *point of presence* (POP). They are also shown in Fig. 6.7.

Signaling system number 7 (SS7) is designed for use in digital networks in conjunction with stored program-controlled (SPC) switches. Labeled messages (packets) are sent between SPC processors in the switches to convey information regarding call setups, call addressing, call routing, and call releasing. The SS7 also can transmit information regarding the type of call, the demanded bandwidth of the channel being set up, call forwarding, calls to 800 numbers (Freephone), the identity of the calling party, etc. Information regarding the billing of calls, including credit-card calls, collect calls, and calls billed to third parties, also can be transmitted by SS7.

The CCITT signaling system number 7 (SS7) is a modular system that specifies functions. These functions can then be implemented in different ways to meet specific requirements, such as different network and user requirements. The functions are divided between a *message-transfer part* (MTP) and *user parts* (UPs). Among specific user parts are the *telephone user part* (TUP), the *data user part* (DUP), the *integrated services digital network user part* (ISDN-UP), and an *operations and maintenance application part*. In this context, the term *user* refers to any functional entity that utilizes the transport capability provided by the message-transfer part (MTP). The MTP provides transport of messages between users. As shown in Fig. 6.5, it consists of three parts covering the open systems interconnection (OSI) layers 1, 2, and 3. The lowest level, corresponding to the OSI layer 1, handles the signaling data link functions and consists of a bidirectional transmission path. Level 2 takes care of the signaling link functions and, in conjunction with level 1, provides a signaling link for the reliable transfer of signaling messages between two directly connected signal points. These messages are called *signal units,* are of three types, and can be of variable length. The level 2 MTP also handles error correction, error monitoring, and flow control. The signaling network functions, level 3, correspond to the lower half of the OSI's network layer (3). They provide the functions and procedures for the transfer of messages between signaling points, which are the nodes of the signaling network. There are two kinds of signaling network functions: signaling message handling and signaling network management.

The *signaling connection control part* (SCCP) of SS7 together with level 3 of the message-transfer part (MTP) makes up the functional equivalent of the OSI network layer, i.e., layer 3. The SCCP enhances the addressing capability of the MTP and provides for four classes of signaling message services.

The SS7 protocol lacks levels corresponding to the OSI layers 4, 5, and 6. Thus the *transaction capabilities application part* (TCAP) directly uses the services of the SCCP. It handles a wide range of protocols and functions referred to as *transactional capabilities* (TCs). The TCAP can, for instance, invoke the substitution of a physical number for an 800 number (Freephone number).

Other parts of the applications layer consist of application service elements (ASEs), and the *operation, maintenance, and administration part* (OMAP), shown in Fig. 6.5.

The *integrated services digital network user part* (ISDN-UP) provided by SS7 is essential for the operation of ISDN. Actually, an ISDN call cannot be established between switches unless the two switches both have access to the same SS7, as discussed in the following section.

While the introduction of SS7 and ISDN occurred relatively rapidly in Europe and Japan, their introduction in the United States has been slow. The reason is that most European telecommunications operations are state owned. It is thus relatively easy for the state to order the introduction of SS7 and ISDN. In the United States and other countries with privately owned telecommunications companies, any such introduction must be economically motivated. Figure 6.8 shows the growth of SS7 and ISDN installations worldwide.

At the end of 1992, the percentage of access lines being connected to switches with SS7 signaling capability ranged from 54 percent for Ameritech (an area including Chicago) and NYNEX (which includes New York City) to 95 percent for Bell Atlantic (covering the mid-Atlantic coast). Some switches have intra-LATA (local access and transport area) SS7 capability only, while others have inter-LATA capability, including nationwide service.

In March of 1991, SS7 had not been installed in New York City yet. Thus switches that offer ISDN on Manhattan could handle ISDN calls between users connected to the same switch only. The widely publicized trial calls between New York, Tokyo, and Paris were not switched.

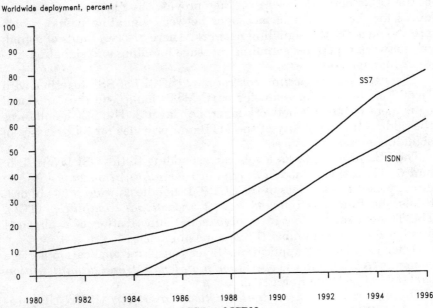

Worldwide deployment, percent

Figure 6.8 Worldwide deployment of SS7 and ISDN.

6.5   B-ISDN Signaling

Figure 6.9 shows an overview of broadband transport and signaling networks. Broadband information is transported over a physical network that can consist of a conventional network and which in the future will be a SONET (or SDH) network. Superimposed on this network will be an asynchronous transfer mode (ATM) network. The ATM network will use the transmission links of the SONET network, but the ATM switches will be physically different from the SONET switches. The figure shows different links between the users and the two networks. Physically, users will have a single link to the combined SONET and ATM networks. Virtually, there will be two separate links.

The B-ISDN network may use different signaling systems, including signaling system number 7 (SS7) or a later version of it. The ATM network will use the signaling system recommended in ITU-T draft Recommendation Q.2931 (former Q.93B). Users will communicate with the physical SONET network using digital signaling system number 2 (DSS2). In addition, SONET can provide message-based signaling through overhead functions mapped into its main bit stream.

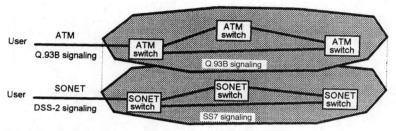

Figure 6.9 Levels of broadband signaling networks.

While the narrowband ISDN signaling systems are designed to control single channels between two users, B-ISDN applications can involve multimedia applications with a call consisting of several different channels. Thus separate channels may be required for voice, data, and video. Multiparty calls also will exist.

Study group XI within the ITU-T is examining a new SS7 application layer structure called the *ISDN service control part* (ISCP). It separates the call control and the bearer control parts of the protocol. Call control messages can take a different network path than the actual bearer channel. A new broadband application part (BAP) is being developed for B-ISDN signaling applications. The current SS7 and DSS1 schemes will be gradually extended to cover multimedia communications and management by the mid-1990s.

References

ITU Telecommunications Standardization Sector, Draft Recommendation Q.93B (1993), "Broadband ISDN user-network interface layer 3 specification for basic call/connection control" October, ITU, Geneva.

Shiraishi, S. (1993), Draft text for Q.2931 (Chapters 1, 2, and 3). International Telecommunications Union, Telecommunication Standardization Sector, Temporary Document 2/11-21, PL/11-24 December, ITU, Geneva.

7

Local Area Networks

7.1 Introduction

Initially, personal computers (PCs) in business and other organizational enterprises were stand-alone devices used for word processing, financial calculations (spreadsheets), etc. Soon the requirement to interconnect them, to connect them to mainframe computers, and to arrange for sharing of peripheral equipment arose. Thus local area networks (LANs) were born. They let users of PCs access data bases in mainframes, share mass storage devices, printers, plotters, etc., and communicate with one another.

Different devices are used to access a LAN depending on the required transfer speed and the protocols involved. Generally, the price of interface devices increases with transfer speed. A practical limit is that the price of the interface device should be a fraction of the price of the device to be connected to the LAN.

The design of local area networks varies with the type of computers connected to them. A LAN interconnecting large supercomputers and mass storage devices is more complex and expensive than an LAN for an office environment with low- and medium-speed data-transfer requirements. A third type of LAN is used in the factory environment to interconnect automated manufacturing and process-control equipment such as programmable controllers, processors, robots, machine vision devices, personal computers and workstations.

In the latest trend, computers and other devices are connected through wireless means. One reason for this is the high cost of rearranging wirebound LANs compared with the very low cost of moving a wireless device.

7.2 Architectures

Typically, a *local area network* (LAN) is a private data communications network that interconnects a number of data terminals and a mainframe computer, printers, or data-base servers. The term *local area* refers to their limited geographic coverage. A small LAN may cover a department on a single floor, while a larger LAN can cover an entire building. Some large LANs cover an entire university campus. The actual maximum transmission distance varies between different types of LANs and ranges from 300 m to 100 km. Systems from major suppliers, such as IBM, offer a maximum distance of 2000 m. Ethernet (IEEE Standard 802.3) has a maximum of 1500 m. The maximum transmission distance depends on the maximum bit rate and the type of physical transmission media.

The operating scheme of practically all LANs is packet switching. In LANs offering so-called connectionless systems, no connections between users are set up. Rather, a transmitting user obtains exclusive access to the medium during transmission. This is accomplished through so-called token passing or collision detection. In the first case, only a terminal in possession of a "token" may transmit; in the latter case, transmission is interrupted if more than one terminal transmits at the same time. The tokens are passed automatically from terminal to terminal. In the case of a collision, attempts to transmit are resumed after a random delay.

The maximum bit rate on the transmission lines depends on the type of protocol used and the transmission medium. For example, Ethernet offers a bit rate of 10 Mbit/s when the transmission medium consists of a LAN coaxial cable. Other versions use LAN twisted-copper pairs or so-called LAN thin coaxial cables, in which case the transmission speed can be reduced to 1 Mbit/s.

Fiber-distributed data interface (FDDI) is a new standard. As the name indicates, the system is based on the use of fiberoptic transmission links. Actually, the recommended architecture consists of two fiberoptic rings with transmission in opposing directions at a bit rate of 100 Mbit/s. Thus FDDI can transmit at 10 times the rate of Ethernet. FDDI uses token passing as its access protocol.

LANs can be arranged in the form of a bus, a ring, a star, a tree, or hybrids of these. The most common are the bus structure used by Ethernet and the ring structure used by IBM, among others. These structures are shown in Fig. 7.1.

Terminals, host computers, printers, etc.—commonly called *data terminal equipment* (DTE)—are connected to the bus by means of a *bus interface unit* (BIU). Information is typically broadcast on the bus and received by the one or several destination DTEs.

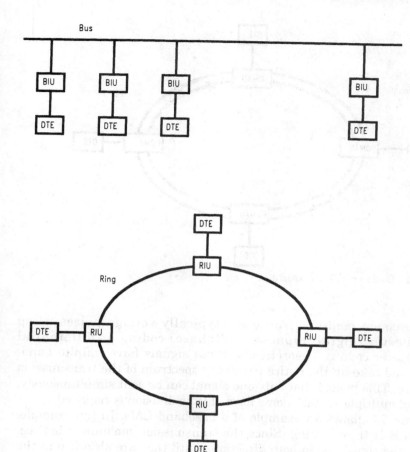

Figure 7.1   Bus and ring LAN structures.

In a ring LAN, the DTEs are connected to the ring through so-called ring interface units (RIUs). The recommended structure for FDDI consists of two rings, as shown in Fig. 7.2. The two rings increase the reliability of the system, since it is unlikely that both rings will be out simultaneously. In high-reliability systems, the two rings are drawn through separate ducts. Typically, the traffic moves in opposite directions on the two rings.

7.3   Transmission Techniques

Two different transmission techniques are used: baseband and broadband. In the *baseband* technique, digital pulses are inserted into the

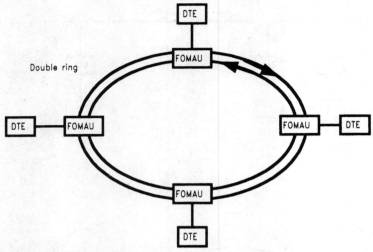

Figure 7.2 Double-ring LAN structure.

transmission medium. These are typically voltage pulses using
Manchester (digital biphase or diphase) coding or differential
Manchester coding. Theoretically, these signals have infinite band-
width and take up the entire frequency spectrum of the transmission
medium. This means that only one signal can be sent simultaneously.
To send multiple signals, some form of time division is required.

Figure 7.3 shows an example of a baseband LAN. In the example,
station B is transmitting. Since the transmission medium is bidirectional, the signals go in both directions until they are absorbed at the
end of the link. In the example it is assumed that the signals from
station B are addressed to station D, to which they are sent.

In the context of LANs, *broadband* refers to a technique that uses
analog signals with frequency-division multiplexing. The frequency
spectrum of the transmission medium is divided into channels, and

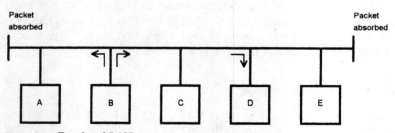

Figure 7.3 Baseband LAN.

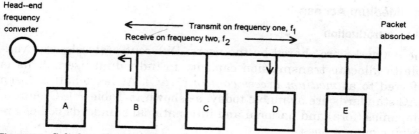

Figure 7.4 Split broadband LAN.

different signals can be sent in the different channels. The transmission is unidirectional, and two channels are required for bidirectional traffic. Figure 7.4 shows an example of a *split broadband LAN*; i.e., signals in the two directions are send on different frequencies, here called f_1 and f_2. To the left in the figure is the head end. It has a frequency converter that, in this case, converts frequency f_1 to f_2. Generally, it converts incoming signals of one frequency to outgoing signals of a different frequency. Using the same example in which station B wants to transmit to station D, signals of f_1 go from station B to the converter, where they are converted to f_2, and from which they go to station D.

A broadband LAN also can use different cables in the two directions and is called a *dual-cable broadband LAN,* as shown in Fig. 7.5. In this case, the same frequency is used in both directions, and no conversion, nor converter, is required. Following the signals in the example where station B is transmitting to station D, here the signals go in the "upper" cable, around the passive head end, and through the "lower cable" to station D.

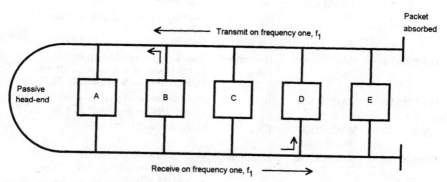

Figure 7.5 Dual-cable broadband LAN.

7.4   Medium Access

7.4.1   Introduction

The networks are shared by the users. Different protocols are available to allocate transmission capacity to individual users. This is referred to as *medium access control* (MAC). At least half a dozen MAC standards are available today, as shown in Table 7.1. Groups of companies (fora) and national and international standards bodies are working on new ones.

7.4.2   Logical link control (LLC)—IEEE Standard 802.2

IEEE Standard 802.2, "Logical Link Control (LLC)," refers to all types of media access and specifies the rules for addressing stations and for controlling the transfer of data between users. The standard is based on the high-level data link control (HDLC) protocol, an ISO standard. It offers three types of service: (1) unacknowledged connectionless service, (2) connection-oriented service, and (3) acknowledged connectionless service.

TABLE 7.1   LAN Standards

Standard	Description	Alternative services
IEEE 802.2	Logical link control (LLC)	Unacknowledged connectionless service Connection-oriented service Acknowledged connectionless service

Standard	Description	Topologies	Physical media
IEEE 802.3	MAC: CSMA/CD	Bus, tree, star	Baseband coaxial at 10 Mbit/s (two versions) Unshielded twisted pair at 1 and 10 Mbit/s Broadband coaxial at 10 Mbit/s
IEEE 802.4	MAC: Token bus	Bus, tree, star	Broadband coaxial at 10 Mbit/s Carrierband at 1, 5, and 10 Mbit/s Optical fiber at 5, 10, and 20 Mbit/s
IEEE 802.5	MAC: Token ring	Ring	Shielded twisted pair at 4 and 16 Mbit/s Unshielded twisted pair at 4 Mbit/s
ANSI X3T9.5	MAC: FDDI token ring	Ring	Optical fiber at 100 Mbit/s
IEEE 805.6	MAC: DQDB	Dual bus	Optical fiber or coaxial at North American signaling level 3 (DS-3, about 44 Mbit/s)

7.4.3 Medium access control (MAC)

Currently, five different standards are available for medium access control (MAC). Four of them are typically referred to by the IEEE committee that put them together, 802.3 through 802.6. The fifth covers MAC for the FDDI token ring.

IEEE Standard 802.3 covers *carrier sensing multiple access with collision detection* (CSMA/CD). In this case, each unit that wants to transmit first senses for the presence of a carrier signal on the transmission medium. If a carrier signal is sensed, it indicates that another unit is transmitting, and the second unit delays its transmission. Owing to transmission delays, it is still possible that two units start to transmit at the same time. If this happens, both cease transmitting under the CSMA/CD system. This standard is used by Ethernet.

Variations of the 802.3 standard are identified by a special notation:

```
<data rate in Mbit/s><signaling method><maximum segment length in
hundreds of meters>
```

In some cases, the letters F and T are used to indicate fiber and twisted pair, respectively, substituted for the segment length number. Examples of such variations are

- 10BASE5
- 10BASE2
- 1BASE5
- 10BASET
- 10BROAD36

In a token ring system, a token is passed along from one station to another around a ring. Only a station that possesses the token can transmit. IBM and FDDI use token rings. IEEE Standard 802.4 specifies a *token bus* standard; i.e., it uses tokens on a bus architecture, including tree and star architectures. Similarly, IEEE Standard 802.5 specifies a *token ring* standard; i.e., tokens are used on a ring architecture.

Fiber distributed data interface (FDDI) uses a token ring and is specified in ANSI Standard X3T9.5. It operates at a speed of 100 Mbit/s and can accommodate up to 500 stations distributed along a total cable distance of 100 km. Efforts have been made to reduce the cost of FDDI by using less expensive fibers and—more important— less expensive optical transmitters and receivers. These efforts have resulted in a low-cost fiber media–dependent physical layer (LCF-PMD) that allows for transmission distances of 500 m. Another alternative, called MMF-PMD, uses 62.5/125-μm multimode fiber (MMF) and can operate over 2 km. FDDI differs from the other standards in that it can support synchronous (time-sensitive) traffic such as voice and video in addition to asynchronous traffic. However, in the synchronous mode, the time when a station has the token for transmission can vary and cause problems for voice and video. In a new version called FDDI-II, this problem is avoided through the addition of an isochronous service mode. Like FDDI, FDDI-II runs at 100 Mbit/s. Today, the name *FDDI* is a misnomer. FDDI is not limited to fiber, it is not always distributed, it covers more than data, and it is more than an interface.

IEEE Standard 802.6 covers a dual-bus system using the *distributed queue dual-bus* (DQDB) coding. It is used mainly for metropolitan area networks and is described later in Chap. 11.

Other protocols include the *polling system,* in which a central unit, typically the host computer, polls one station after another of those connected to the system, basically asking it to transmit if it has something to transmit.

7.5 High-Speed LANs

With current and future demands for bandwidth, not even FDDI will suffice. Among attempts for higher-bit-rate LAN protocols are HIPPI, fibre channel, FDDI follow-on LAN (FFOL), and local ATM (LATM).

7.5.1 High-performance parallel interface (HIPPI)

The Los Alamos National Laboratory suggested in 1987 that a higher-speed LAN was required to interconnect supercomputers. As a result, the high-performance parallel interface (HIPPI) was developed. It operates at 800 Mbit/s on one interface cable and at 1600 Mbit/s on two cables. It provides transfer in one direction, and two back-to-back HIPPIs are required for duplex transmission. The cables are twisted-pair copper wires and are limited to 25 m in length. Thirty-two-bit parallel words are transmitted on the 800-Mbit/s HIPPI and 64-bit words on the 1600-Mbit/s HIPPI. The working distance of HIPPI can be extended beyond 25 m to several kilometers through the use of special flow-control mechanisms. There is also a serial HIPPI, in which the parallel cable is replaced by a single metallic or fiber cable. It uses a 20-bit/24-bit coding scheme resulting in a 1.2-GBd serial signal and can operate over a distance of up to 10 km using 9-μm single-mode fiber. Coaxial copper cables can be used for distances up to 36 m.

7.5.2 Fibre channel

Fibre channel (FC) supports four speeds with data transfer rates of 100, 200, 400, and 800 Mbit/s, corresponding to 132- , 266- , 531- , and 1062.5-MBd serial signaling rates. It can operate on single-mode optical fiber, multimode fiber, and even copper coaxial cable for short distances. Using 1300-nm lasers and single-mode fiber, FC can operate over distances of 10 km.

7.5.3 FDDI follow-on LAN (FFOL)

FFOL is designed to support interfaces in wide area networks and is described later in Chap. 11. It will operate at up to 2.5 Gbit/s on links with a maximum distance of 10 km and on a maximum network distance of 100 km.

7.5.4 Asynchronous transfer mode (ATM) hub

An ATM switch in the hub of a star-formed LAN will provide very high-speed LANs. While an FDDI LAN operates at 100 Mbit/s, that bandwidth is shared by all users connected to that LAN. In the case of an ATM hub, each user can access the ATM switch at its maximum access rate, be it 155 Mbit/s, 622 Mbit/s, or higher.

References

Jain, Raj (1993), "FDDI: Current issues and future plans," *IEEE Communications Magazine,* vol. 31, no. 9, September, pp. 98–105.
Stalling, William (ed.) (1993), *Advances in Local and Metropolitan Area Networks,* IEEE Computer Society Press, Los Alamitos, Calif.
Tolmie, Don E. (1992), "Gigabit networking," *IEEE LTS,* May, pp. 28–36.

8

Broadband
Digital Trunks

8.1 Introduction

The introduction of digital technology in telecommunications was begun in the transmission area. Originally, a set of asynchronous digital signal levels was introduced. A higher digital signal (DS) level was obtained as a multiple of lower levels plus additional bits to accommodate differences in the asynchronous transmission. These extra bits allowed for stuffing (and removal) of bits to smooth the differences in the bit rates of the individual lower-level channels. This method is called *plesiochronous digital hierarchy* (PDH) from the Greek word *pleasio* meaning "almost." Table 8.1 shows the digital signal levels in the North American system (and in Japan), and Table 8.2 shows those recommended by CCITT (now ITU, Telecommunications Sector) and used outside of North America and Japan. As mentioned previously, the digital signal levels according to CCITT Recommendation G.702 are called E-1 through E-4 in Europe.

The PDH method has some inherent problems. When lower-level signals are added to or dropped from a higher-level signal, the signals at each level have to be multiplexed and demultiplexed, respectively. It is also difficult to manage a PDH network.

8.2 Synchronous Optical Network (SONET)

The introduction of digital fiberoptic transmission facilities with potential bit rates in the terabit per second range forced the consideration of new digital transmission techniques. In particular, the need

TABLE 8.1   North American Plesiochronous Digital Signal
Level Hierarchy

Digital signal level	Bit rate (kbit/s)	Number of equivalent voice-frequency channels
DS-0	64	1
DS-1	1,544	24
DS-1C	3,152	48
DS-2	6,312	96
DS-3	44,378	672

for multiplexing and demultiplexing at all levels arose. Thus a synchronous hierarchy called *SONET* (for *synchronous optical network*) was developed in the United States by the American National Standards Institute (ANSI), with ANSI Standards T1, 105 and T1, 106 issued in 1988. During that same year, the CCITT (now ITU-T) issued Recommendation G.707 for the similar synchronous digital hierarchy (SDH). For political reasons, the basic bit rate for SONET is 51.84 Mbit/s and that for SDH is 155.52 Mbit/s. All higher levels of bit rates are integer multiples of these basic rates. The basic rate for SDH is exactly three times that of SONET. Table 8.3 shows these rates as well as the payload rates, i.e., with the overheads deducted. One entire frame is transmitted every 125 µs. This value corresponds to the sampling rate of 8000 per second used for voice digitizing, since 1 s÷8000 = 125 µs. SONET uses the term *synchronous transport signal* (STS) and SDH uses *synchronous transport module* (STM) for the electric signals of the different levels. In SONET, the corresponding optical levels are called *optical carrier* (OC).

A SONET STS-1 frame can be seen as a block with nine rows containing 90 bytes (720 bits) each, for a total of 810 bytes, or 6480 bits,

TABLE 8.2   CCITT (ITU-T) Plesiochronious Digital Signal
Level Hierarchy

Digital signal level	Bit rate (kbit/s)	Number of equivalent voice-frequency channels and signaling channels
1st	2,048	30 + 2
2d	8,448	120 + 8
3d	34,368	480 + 32
4th	139,264	1920 + 128

SOURCE: CCITT Recommendation G.702.

TABLE 8.3 Signaling Levels in SDH and SONET

Synchronous transport module	Synchronous transport signal	Optical carrier	Nominal rate, Mbit/s	Payload rate, Mbit/s
	1	1	51.84	50.112
1	3	3	155.52	150.336
3	9	9	466.56	451.008
4	12	12	622.08	601.344
6	18	18	933.12	902.016
8	24	24	1244.16	1202.688
12	36	36	1866.24	1804.032
16	48	48	2488.32	2405.376
32	96	96	4976.64	4810.176
64	192	192	9953.28	9620.928

as shown in Fig. 8.1. The first three columns contain the *transport overhead*. Of these columns the first three rows (9 bytes) are called the *section overhead*, and the following six rows (18 bytes) are called the *line overhead*. Here the term *line* refers to the transmission medium between terminals that originate and terminate optical carrier (OC) signals, as shown in Fig. 8.2. Similarly, *section* refers to the portion of a transmission facility between these terminals and a regenerator or between two regenerators. Bytes 4 through 90 in each of the nine rows are available for *payload* information. In order to maintain the relationship with the 8000-per-second sampling rate (of voice), each frame has a length of 125 µs, equal to 1 s ÷ 8000. The *term path* in Fig. 8.2 refers to the connection between two pieces of terminal equipment.

The SONET overheads provide message-based channels for alarms, maintenance, control, monitoring, administering, signaling, and other communication needs. These messages can be generated internally or externally or be specified by equipment manufacturers. The availabil-

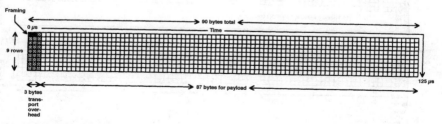

Figure 8.1 SONET STS-1 frame.

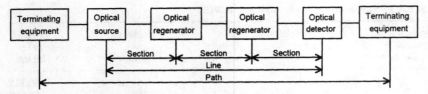

Figure 8.2   SONET section, line, and path.

ity of these message channels greatly enhances the usefulness of SONET (and SDH).

Frames are contained in continuous serial transmissions of bits. Special framing bits are used to indicate the beginning of a frame (and the end of the preceding frame). These framing bits are contained in the first 2 bytes of the first row, as indicated in Fig. 8.1.

In order to maintain flexibility and to accommodate bit streams that are delayed, subject to jitter, or out of synchronization, the mapping of information into the SONET STS frames does not necessarily have to start in the first payload frame (the fourth byte in the first row of the STS-1 frame). Rather, a payload frame can start anywhere in the STS frame. Information regarding that starting point is contained in a *pointer*. These pointers have a fixed position in the STS frame and are specifically located in the first 3 bytes of the fourth row. This location is identical to the first row of the line overhead. A number in the pointer indicates the number of bytes in the payload envelope counted from the end of the pointer where the payload information starts. In the example in Fig. 8.3, the pointer would have number 8 because there are 8 bytes between the end of the pointer and the start of the payload frame. Besides the use of point-

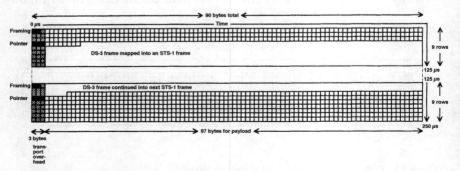

Figure 8.3   DS-3 frame mapped into SONET STS-1 frame.

ers, other approaches let the payload frames start in a fixed position and take care of deviations in the arrival time of frames through bit or byte stuffing.

8.3   Network Layout

Among the major demands on a trunk network are reliability and survivability. The network should function even though one or several links and/or nodes fail. This can be accomplished by a mesh network and/or through one-for-one protective switching. A more economic approach is the combination of ring networks where each ring consists of a double ring with traffic in opposite directions on each ring as shown in Fig. 8.4. It shows two rings connecting nodes 1, 2, 3, and 4 with one ring having clockwise traffic and the other counterclockwise traffic. The same information stream is sent in both directions and the destination picks up the one with the best quality. This arrangement is also called a unidirectional path switching ring (UPSR). If, for example, the link between nodes 3 and 4 is broken, traffic from node 3 can reach node 4 by way of nodes 2 and 1. To keep

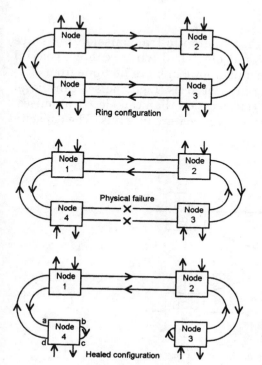

Figure 8.4 Ring configurations: original, broken, healed.

traffic moving between the four nodes, traffic that normally would leave port b of node 4 for node 3 is redirected to port c and leaves node 4 at port d. The rerouting through ports b and c is typically done inside node 4. An alternative to UPSR is bidirectional line switched ring (BLSR) as per Bellcore TA-NWT-001230.

The equipment in the nodes in Figure 8.4 typically consists of add-drop multiplexors (ADMs). As the name implies, an ADM permits the adding and dropping of traffic on a ring. A more expensive alternative is to use two terminal multiplexors back-to-back.

A digital cross-connect system (DCS) provides another way of ensuring reliability and survivability. A DCS offers a centralized point of consolidating. rerouting, and testing of traffic. DCSs have been available for narrowband traffic for a long time and are becoming available for broadband traffic. Figure 8.5 shows four nodes with DCSs.

The SONET standard allows for the interconnection of different types of devices. Thus digital cross-connect devices and add and drop multiplexors can be interconnected to form a hybrid network as shown in Fig. 8.6. Here the ADMs offer simplicity, and the DCSs offer flexibility.

8.4 Fiberoptic Trunks

There are some indications that fiberoptic transmission links are more reliable than conventional electrical wires, cables, coaxial cables, etc. Light pulses inside a fiberoptic link are immune from electromagnetic and other outside interferences. Light pulses can travel over longer distances without regeneration than can electrical pulses. This reduces the required number of repeaters in a link. The capacity

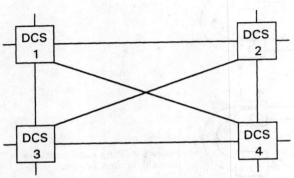

Figure 8.5 Digital cross-connect system.

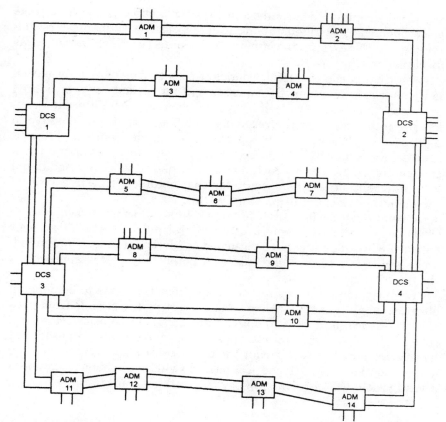

Figure 8.6 Hybrid digital cross-connect system and ring configuration.

of fiberoptic links is much higher than that of wires, cables, and even coaxial cables. A single optical fiber can carry pulses at rates of up to tens of terabits per second (Tbit/s, i.e., 10^{12} bit/s).

The traffic between North America and Europe has grown at an average annual rate of about 24 percent over the last 30 years. Other transoceanic traffic is also growing. Fiberoptic links are also used to meet this growing demand for transoceanic traffic. The first transatlantic fiberoptic cable, the TAT-8, was installed between North America and Europe in 1988. Other such cables have followed in the Mediterranean and the Pacific Ocean.

So far, fiberoptic links have been used mainly between large switching nodes and have not generally been used as access lines. In the cases of large PBXs connected to central offices, fiberoptic links are economically justified and are used. Operating companies are contemplating the extension of fiberoptic links to customer premises, as discussed in Chap. 4.

Even though the capacity of optical fibers to transport information is very high, means of further increasing that capacity are considered and used. One way is to feed light from sources with different wavelengths into a fiber. At the receiving end, filters that are sensitive to light of a single wavelength separate the signals from the different sources. Each pair of light source and receiver can then carry its own specific information; i.e., each source can be individually modulated with information that is specific for that channel. This is a form of wavelength-division multiplexing (WDM).

As mentioned earlier, each fiberoptic link is terminated in electric circuits. In order to regenerate the digital pulses transmitted over optical fibers, they are to be returned to electrical pulses, amplified, and again turned into light pulses. A potentially better approach is the use of optical amplifiers.

Currently, circuit switching of optical fiber links has to be done on electrical signals. Again, this means converting the signals from optical to electrical and back to optical, an inefficient solution. However, several different types of optical switches have been proposed and tried. Such a switch will permit the direct controlled circuit switching of light signals. Similarly, optical packet switching schemes are under investigation. Optical switching is discussed in Chap. 10.

References

Bellcore Technical Advisory TA-NWT-001230 (April 1993), "SONET Bidirectional Line Switched Ring Equipment Generic Criteria," Issue 3, Bellcore, Livingston, N.J.

Bellcore Technical Reference TR-NWT-000496 (September 1991), "SONET ADM generic criteria: a unidirectional, dual fed, path protection switched, self-healing ring implementation," Supplement 1, Bellcore, Livingston, N.J.

Bellcore Technical Reference TR-NWT-000253 (December 1991), "Synchronous optical network (SONET) transport systems: common generic criteria," Issue 2, Bellcore, Livingston, N.J.

Ching, Yau-Chau, and H. Sabit Say (1993), "SONET Implementation," *IEEE Communications Magazine*, vol. 31, no. 9, September, pp. 34–40.

Omidyar, Cambyse Guy, and Anne Aldridge (1993), "Introduction to SDH/SONET," *IEEE Communications Magazine*, vol. 31, no. 9, September, pp. 30–33.

9

Digital Satellite Communications

9.1 Introduction

Satellites have been used since the 1960s to form transmission links across oceans and continents. Figure 9.1 shows a typical arrangement. A satellite is orbiting the earth at a high altitude. If that altitude is 35,900 km above the equator, the angular velocity of the satellite becomes the same as that of the earth, and the satellite appears to be stationary above a certain point on the earth. The satellite is said to be in a *geostationary orbit*. Signals are transmitted from an earth station to the satellite on a so-called *uplink*. A *transponder* in

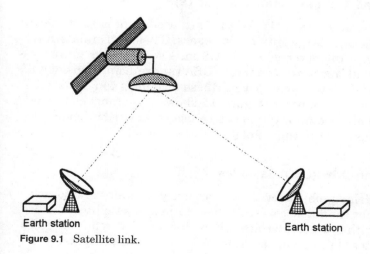

Earth station

Earth station

Figure 9.1 Satellite link.

the satellite amplifies the signals and converts them to another frequency for transmission back to another earth station on a *downlink*. The uplinks typically use the band of 5.9 to 6.4 GHz and the downlinks 3.7 to 4.2 GHz. In addition, the 12- and 14-GHz bands are used for downlinks and uplinks, respectively. Experiments with 30-GHz uplinks and 20-GHz downlinks are being conducted (Levitt, 1992).

Satellites have to be stabilized so that their antennas are directed to a specific point on earth. This is accomplished through the use of the gyroscopic effect; either the entire satellite or a section of it rotates, or gyroscopes are built into the satellite. The microwaves used for transmission are subject to atmospheric interference, such as rain. When the sun is positioned at the extension of the line of transmission between the satellite and a receiving earth station, transmission problems are also encountered. This means that satellite transmissions are not always reliable. The positioning and orientation of satellites in orbit are accomplished through the ejection of gases through jets. The supply of these gases determines the active life of a satellite, which typically is 7 to 10 years.

Geostationary satellites perform well for television broadcasting but are less suitable for two-way transmissions due to the transmission delays. A round trip for the signals takes about half a second, and this is annoying for users. The problem doubles if more than one satellite hop is used, and this is the main reason why more than one hop is avoided.

Because a satellite in a geostationary orbit seems to be stationary in the sky, antennas can easily be permanently directed toward the satellite.

9.2   Very Small Aperture Terminals (VSATs)

Very small aperture terminals (VSATs) are used to broadcast television programs to a large number of users. These terminals have antennas with a diameter of 1.2 or 1.8 m. Even smaller units, so-called ultrasmall aperture terminals (USATs) are being considered. In addition to television broadcasting, these terminals can be used for data transmission from users. Figure 9.2 shows an arrangement with VSAT. A normal-sized hub station can be used for distribution of television programs or for reception of data transmissions.

9.3   Low-Earth-Orbit (LEO) Satellites

To avoid the delay problems with geostationary satellites and to cover polar areas that cannot be covered by geostationary satellites, the use of satellites with low earth orbits (LEOs) has been introduced. Since these satellites do not appear stationary to earth stations, some

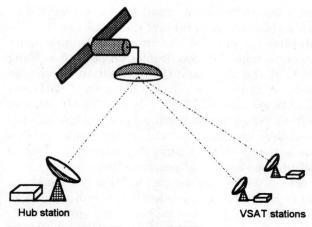

Figure 9.2 Satellite links with VSATs.

arrangement by which the antennas follow the satellites is required. LEO systems have been suggested for mobile and personal communications networks (PCNs).

9.4   Frequency Utilization and Coding

Frequencies for satellite transmissions are finite, and the demand is increasing. There are several ways to increase the usage of available frequencies. One is to use polarized transmissions, sending different signals in the two polar planes. In addition, the orbit suitable for geostationary satellites is getting crowded, and few new positions are available. Satellites whose orbits are sufficiently separated can reuse the same frequency. By using directional antennas, the transmission beams can be sent to different spots on earth.

Originally, satellite transmissions were analog. Today, digital transmission is also available. Some systems use quaternary phase-shift keying (QPSK) of a carrier and can transmit digital signals at rates of up to 64 Mbit/s. Using time-division multiple access (TDMA), several different streams of digital signals can be sent on the same satellite transmission link. Another technique is the use of spread spectrum with code-division multiple access (CDMA). Here, the *code* refers to the coding of the spectrum signals.

9.5   Potential Applications

The Japanese are using satellite communications to make narrowband ISDN available to remote users (Kato et al., 1992). Petr et al.

(1992) have suggested a bandwidth-on-demand satellite communications network. They talk about transfer rates of up to 768 kbit/s.

Several uses of satellites for broadband transmissions have been suggested. One is for connection of remote local area networks (Yang et al., 1992). Aghvami et al. (1992) discuss the modulation techniques required for transmission of synchronous digital hierarchy (SDH) and synchronous optical network (SONET) signals at 155 Mbit/s and higher over satellite links. They suggest using 16-ary QAM and 64-ary QAM transmissions over standard 54- and 36-MHz analog satellite links. The purpose is to use satellite links as a backup for terrestrial SDH and SONET links or as an alternative to them.

However, owing to the efficiency and low cost of fiberoptic long-haul transmission links, they have practically taken over from the use of satellite links. In the future, satellites are expected to be used mainly for mobile communications and personal communications networks and to reach remote areas.

References

Aghvami, A. Hamid, Orhan Gemikonakli, and Shuzo Kato (1992), "Transmission of SDH signals through future satellite channels using high-level modulation techniques," *IEEE Journal of Selected Areas in Communications*, vol. 10, no. 6, August, pp. 1030–1036.

Kato, Shuzo, Masahiro Morikura, Shuji Kubota et al. (1992), "A TDMA satellite communication system for ISDN services," *IEEE Journal on Selected Areas in Communications*, vol. 10, no. 2, February, pp. 456–464.

Levitt, Barry K. (1992), "Rain compensation algorithm for ACTS mobile terminal," *IEEE Journal on Selected Areas in Communications*, vol. 10, no. 2, February, p. 358.

Petr, David W., K. M. S. Murthy, Victor S. Frost, and Lyn A. Neir (1992), "Modeling and simulation of the resource allocation process in a bandwidth-on-demand satellite communications network," *IEEE Journal of Selected Areas in Communications*, vol. 10, no. 2, February, pp. 465–477.

Yang, Oliver W. W., Xiao-Xiong Yao, and K. M. S. Murthy (1992), "Modeling and performance analysis of file transfer in a satellite wide area network," *IEEE Journal of Selected Areas in Communications*, vol. 10, no. 2, February, pp. 428–436.

10

Digital
Broadband Switching

10.1 Introduction

An important feature in telecommunications is the possibility of connecting different users and to switch between users. It differs from fixed point-to-point connections that remain set up for a long time or even forever. Switching makes it possible to talk to our butcher in one call and to friends and business contacts in others.

The first telephone switches were manual, and an operator established a connection between two users by connecting two jacks with a cord. Soon automatic switches were introduced. The industry has gone through hundreds of different types of switches. Over the years, they have become more automatic, larger, and faster.

Until now telecommunications switches have switched electric circuits, i.e., circuits that carry electric signals. The advent of fiberoptic links between switches has changed this situation. In the future it will be desirable to switch signals that consist of light. We will talk about *optical switching* or—because radiant energy consists of photons—*photonic switching*.

The fabric of a switch can take many different shapes, as indicated in Fig. 10.1. For close to a century, practically all telecommunications switches were based on a fabric using *space division*. The procedure is called *circuit switching* or *space switching,* and the switches are called a *circuit switch* or a *space switch*. Circuit switching can be used to switch circuits carrying either analog or digital signals.

Circuit switches can be divided into those with a *single path* and those with *multiple paths*. Here, *path* refers to the available route between the switch's input and output. Usually only one of the avail-

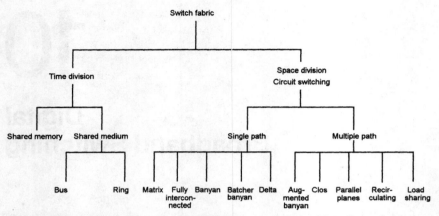

Figure 10.1 Switching fabrics. (*Source: IEEE Communications Magazine, vol. 30, no. 4, April 1992, page 92, Fig. 2.*)

able paths is actually used in a multiple-path switch fabric, but several can be used to obtain redundancy.

In the mid-1970s, switches based on *time division* came into use for switching of digital signals, i.e., signals consisting of pulses representing ones and zeros. Such a switch is called a *time switch*.

A basic type of circuit switch for connecting different circuits is shown in Fig. 10.2. Three lines called *A1, A2,* and *A3* come in from the left. Through two crossbar switches and a couple of trunks, *T1* and *T2*, they can be connected to any of the three lines to the right, *B1, B2,* or *B3*. The two switches are controlled by equipment in the box marked "common control." As an example, the common control can activate the left switch in such a way that line *A2* is connected to trunk *T1*. Similarly, it connects *T1* with line *B3*, thus establishing a

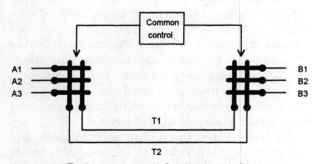

Figure 10.2 Basic arrangement for circuit switching.

connection between lines *A2* and *B3*. We can extend the example by connecting line *A1* with line *B1* over trunk *T2*. Now, if we want to connect lines *A3* and *B2*, it cannot be done because there is no free trunk. Both *T1* and *T2* are busy. This situation is called *blocking*. If *A3* wants a connection with *B3*, it cannot be established because *B3* is *busy* with *A2*. In this example, the circuits *A1, A2, B1,* and *B3* are connected through circuit switching.

In a time-division fabric, digital signals from several different sources (say customers) are interwoven and transmitted on a single circuit. The upper part of Fig. 10.3 shows three sequences of digital signals representing 101, 110, and 111, respectively, being interwoven or multiplexed into one sequence (101110111). In order to accommodate the three individual sequences on one circuit within the same time frame, the number of pulses per time unit has to be three times as large compared with the original sequences.

A digital pulse is usually called a *bit* because it represents a binary digit, a one or a zero. The speed with which digital information (i.e., bits) is transmitted over a circuit is usually measured in bits per sec-

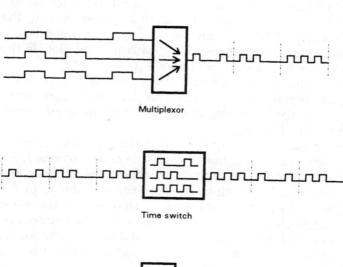

Multiplexor

Time switch

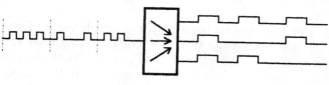

Demultiplexor

Figure 10.3 Multiplexor, time switch, and demultiplexor.

ond, abbreviated bit/s. A thousand bits is called a *kilobit,* abbreviated kbit, according to international standard; a million bits, a megabit (Mbit); a billion (thousand million) bits, a gigabit (Gbit); and a billion billion bits, a terabit (Tbit). (Note that *billion* is used in the American way, meaning 10^9, not 10^{12} as in England.)

In the preceding example, we had three digital sequences that were multiplexed into one sequence. If the bit rate of the three individual sequences were 1 kbit/s each, the bit rate of the multiplexed sequence would have to be 3 kbit/s in order to accommodate all three sequences within the same time frame.

In a multiplexed digital sequence, each of the original sequences is said to occupy a time slot. In our example with three original sequences, there are three time slots in the multiplexed sequence. A more common multiplexed sequence is called DS-1 (digital signal one) and is typically carried on a North American T1 carrier. The DS-1 signal consists of 24 original digital sequences and thus has 24 time slots. A similar European system, called E-1, has 32 time slots.

The second row in Fig. 10.3 shows the multiplexed digital sequence being fed to a time switch of the *shared-memory type.* This time switch stores the incoming digital sequence in a memory, one time slot at a time. In other words, the original digital sequences are stored separately. Under control of the switching mechanism, the time slots can be read out from the memory in any order desired. As a result, the original digital sequences occupy different time slots in the output; i.e., they have been switched in time.

After demultiplexing, as shown in the bottom row of Fig. 10.3, the individual original digital sequences arrive on circuits that are different from the ones in the beginning of the first row. In other words, they have been switched in space.

Even though the main switching element in the preceding example is a shared memory acting as a time switch, the entire arrangement acts as a space switch in the sense that the signals are switched and directed to specific physical circuits. The multiplexor shown in the first row usually always allocates the same time slot to the same physical input circuit. Similarly, the demultiplexor in the bottom row always allocates information in the same time slot to a specific output circuit. Space and time switches can be combined in several ways to form larger switches.

Space switches can be substituted for the multiplexor and the demultiplexor in the preceding example. Thus the incoming lines can be switched in space before they reach the time switch. Similarly, the output from the time switch can be switched in space onto the outgoing circuits. This combination is called a *space-time-space* (STS) *switch.* Other combinations exist, such as TST, STSTS, etc.

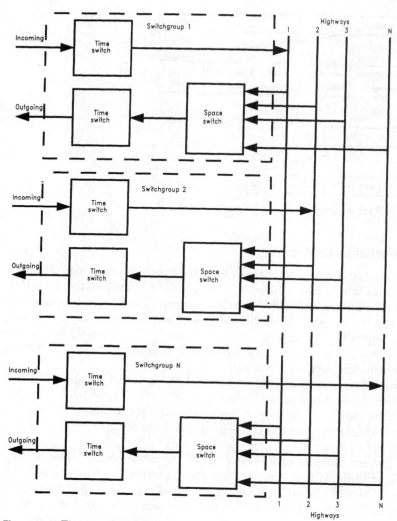

Figure 10.4 Time-space-time switch.

The most common digital switches are of the time-space-time variety. Figure 10.4 shows a typical arrangement of such a switch. It consists of N switch groups, each with one time switch for incoming traffic and one for outgoing. The incoming time switch of each switch group is connected directly to a highway. All the N highways are connected to outgoing time switches by means of a space switch in each switch group.

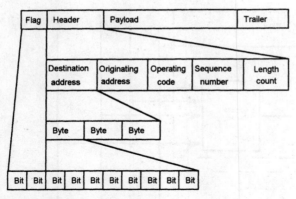

Figure 10.5 Packet format.

10.2 Packet Switching

In *packet switching,* information is transmitted in the form of packets, each of which has a header, a *payload,* and, optionally, a *trailer,*
as shown in Fig. 10.5. The *header* contains an address indicating
where the packet should be directed. A header also can contain information regarding its origin, the type of packet, its priority level, etc.
If there is a *trailer,* it usually contains code for error detection and,
again optionally, error correction.

A *frame* is a group of bits or bytes that needs to be identified for
synchronization purposes. A frame typically starts (and/or ends) with
a *flag* that indicates the beginning or the end of a frame. Packets can
be of a variable length or of a fixed length. In the latter case, a packet
is often called a *cell.*

The general outline of a *packet switching network* is shown in Fig.
10.6. Terminals, which can be data, voice, picture, or video terminals,
are connected to a switching *node* through access lines. As in the case
of a circuit-switched network, the nodes are interconnected with
trunks.

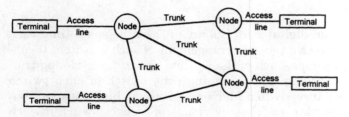

Figure 10.6 Packet switching network.

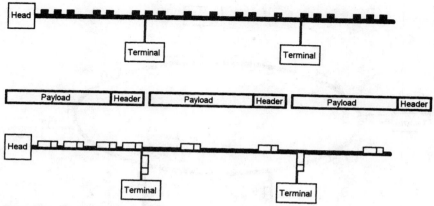

Figure 10.7 Bus switch.

A time-division switching arrangement can be used for packet switching. Typically, it consists of a transmission channel in the form of a ring or a bus from which signals are tapped according to some kind of convention. Groups of signals can be counted and routed to different outlets depending on their position in a sequence of signals. In other approaches, each group of signals contains an address that indicates where to route the group. In Fig. 10.7, digital signals (bits) are traveling down a bus past two terminals. These bits are grouped in a header and a payload, as shown in the middle of the figure. Each header contains an address according to which the packet is directed to a specific terminal. Instead of a bus, the medium can be a ring, as shown in Fig. 10.8.

The incoming circuits to a digital switch can range from subscriber lines with bit rates of 56 to 64 kbit/s to conventional (T-3) long distance lines with bit rates of 44 Mbit/s. Fiberoptic links have higher bit rates, currently about 145 to 155 Mbit/s. Future optical links will have bit rates of 1 Tbit/s and higher. These high bit rates cause problems for conventional digital switches. Present space switches cannot operate fast enough to switch bit streams in the upper gigabit and terabit ranges. Nor can the memories in the time switches read information in and out at such high speeds. One possible solution to this problem is to use optical (photonic) switches.

10.3   Switching in Asynchronous Transfer Mode (ATM)

Asynchronous transfer mode (ATM) has been accepted as the preferred mode for handling of broadband traffic of different kinds. It is

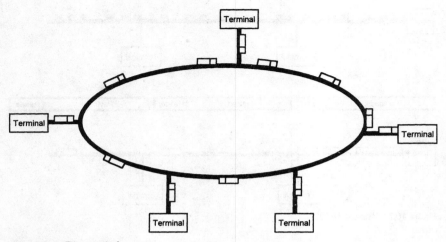

Figure 10.8 Ring switch.

suitable for bursty as well as continuous traffic, including voice, data, pictures, and video. It is a variety of packet switching with fixed-length packets called *cells*, as stated earlier. The cells are transported over *virtual paths* and *channels* that are set up for each connection after negotiation between the user and the network, thus providing a *connection-oriented transport service*, as shown in Fig. 10.9. The fact that the cells have a fixed length and that they traverse a fixed virtual channel means that the delay of different cells is basically the same and that the cells arrive at the destination in *sequence* and without *delay variation*. This is the reason why ATM is suitable for voice and video, two transmission forms that cannot tolerate variations in time for the arrival of information.

Each ATM cell consists of a 5-byte (octet) header and a 53-byte (octet) payload. Figure 10.10 shows the layout of an ATM cell at the user-network interface (UNI), and Fig. 10.11 shows the layout of a

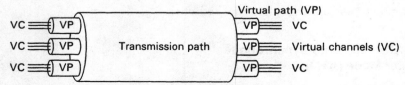

Figure 10.9 Virtual channels, virtual paths, and the transmission path. (*Source: CCITT Recommendation 1.311.*)

bit: 7 6 5 4 3 2 1 0

byte

Generic flow control	Virtual path identifier (VPI)	1
Virtual path identifier (VPI)	Virtual channel identifier (VCI)	2
Virtual channel identifier (VCI)		3
Virtual channel identifier (VCI)	Payload type identifier	Reserved	Cell loss priority	4
Header error check		5
Payload (48 bytes)		6
	53

Figure 10.10 ATM cell at user-network interface (UNI).

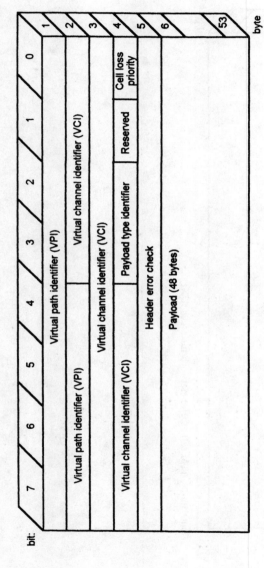

Figure 10.11 ATM cell at network-node interface (NNI).

cell at the network-node interface (NNI). The two cell layouts differ in that the UNI has a field with generic flow control, which is lacking in the NNI. Use of the 4-bit generic flow control field depends on the equipment in operation. The other fields include virtual path identifiers (VPIs) and virtual channel identifiers (VCIs). The VPI and VCI fields identify the virtual path and channel requested by the cell. Both types of cells have a 3-bit payload-type identifier (PTI) that is used to describe one of eight different types of payloads. The 1-bit cell loss priority (CLP) field identifies one of two priorities requested by the user and is translated into two different levels of expected cell loss ratios. A low cell loss ratio is marked by a "1" and a high loss ratio by a "0." The header error check (HEC) field is used to detect or correct errors in the header information depending on the system setup and requirements. The HEC byte is also used for cell delineation. In addition, one bit is reserved for future use. In some versions this reserved bit is already allocated to the payload-type identifier.

The *transmission path* refers to a physical link between two points. On the link, a set of virtual paths is set up. On these virtual paths, a set of virtual channels is set up. These virtual channels carry the ATM cells. Switching between paths and channels is provided at the nodes at the ends of each transmission path.

Virtual channels are the facilities that carry broadband information between users. They usually establish connections between two users. However, links also can be established between several users. In order to save facilities, virtual channels that need to be established between two points are bundled together to form a *virtual path*.

Establishing connections can be accomplished on three levels: The transmission path can be changed through switching, the virtual paths can be changed through switching, and the virtual channels can be changed through switching. Figure 10.12 shows these principles. Each channel and path are identified by a virtual channel identifier (VCI) and by a virtual path identifier (VPI). In the bottom of the figure, the path VPI 4 containing two channels, VCI 1 and VCI 2, is switched to path VPI 5, which contains the same two channels, VCI 1 and VCI 2, through a VP switch. Above that in the figure, virtual path VPI 1 contains two other virtual channels, also called VCI 1 and VCI 2. This path VPI1 is split and switched to two virtual paths, VPI 2 and VPI 3, through the same VP switch. In addition, the two virtual channels, VCI 1 and VCI 2, are switched in a VC switch so that channel VCI 1 leaves as channel VCI 3 and VCI 2 leaves as channel VCI 4.

Cells traversing an ATM switch architecture are assigned to a specific transmission path, virtual path, and virtual channel according to information in the cell's header. Since ATM is connection oriented,

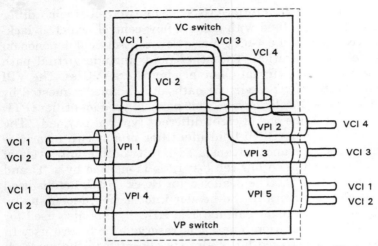

Figure 10.12 Principle of switches for virtual paths and channels. (*Source: CCITT Recommendation 1.311.*)

there is no need for a global address in the cell header. The only addressing required identifies the virtual path and the virtual channel within the transmission path to which the cell belongs. This is accomplished through a virtual path identifier (VPI) and a virtual channel identifier (VCI), respectively. A transmission path and its associated switch ports typically handle traffic with bit rates of 51, 155, and 622 Mbit/s and higher. With a standard cell size of 53 octets, over 120,000, 360,000, and 1,440,000 cells, respectively, will appear at a port every second. Thus there is not time to read and act on more information than is absolutely necessary.

If cells from different inputs compete for the same output port, *blocking* can occur. Depending of the type of switching architecture, blocking also can occur under other circumstances. Besides at the output port, blocking can occur in the middle and the input of certain architectures. Blocking results in the loss of cells. Three means of avoiding these drawbacks are available: *buffer* the information, *discard* some cells, and/or exercise *backpressure*, i.e., prevent further cells from being sent to the congestion point.

Discarding cells should be avoided in the case of data and pictures. Using backpressure delays further transmission and is not recommended for voice and video. Buffering is the most common remedy. Buffers also can take care of the problem with cell delays. In this case, information should be released from the buffers in a timely fashion. Cells can be buffered at the input of a switching element, at the

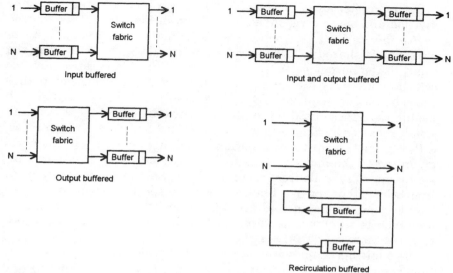

Figure 10.13 Buffering arrangements.

output, within the element, or at combinations of these locations. Figure 10.13 shows several buffering arrangements.

Buffering at the input of a switch results in a performance that is roughly 58 percent of that obtainable with output buffering, according to simulations reported by Karol et al. (1987). To avoid blocking in the case of an output-buffered switch, every output port must be able to accept cells from all input ports at the same time. This is unreasonable to accomplish in a relatively large switch. In the case of a so-called knockout switch, Yeh et al. (1987) state that a cell loss rate of 10^{-6} can be expected if each output buffer can accept eight cells within a single time slot and the traffic load is 90 percent of uniform random traffic. Otherwise, cells will be discarded or other measures will have to be taken.

The two approaches can be combined, and buffers can be placed both before and after the switch fabric. Rather than discarding cells that cannot be handled by the output buffer in a time slot, these cells are retained in the input buffers. This is a design favored for many large switches. Recirculation buffering is a fourth approach. Here cells that cannot be handled at an output port are recirculated to the input side of the switch fabric via recirculation buffers. A problem with this approach is that cells may get out of sequence and special means to avoid this may be required.

Switching fabrics can be classified into two general categories: time division and space division, as was shown in Fig. 10.1. *Time-division switch fabrics* use a common transmission path, a medium or a memory, that is shared by all the cells flowing along the path from the input to the output ports. It is called *time division* because any two cells do not occupy the same spot in time on the transmission path— the cells flow in sequence. Typical examples of fabrics with a shared medium are those used for local area networks and those which consist of a bus or a ring, as shown in Figs. 10.7 and 10.8. In both cases, cells are picked up by destination terminals or other output ports according to the address contained in their headers. In fabrics with shared memory, cells from input ports are read into a large memory, from which they are read out and sent to specific output ports in a similar manner, i.e., according to their designated destination. This approach is similar to that of the time slot interchanges in digital telephone and data switches.

The limiting factor with these types of time-division switches is that their maximum capacity in throughput is determined by the maximum bandwidth of the medium or the maximum storage capacity of the memory. These upper limits cannot be exceeded. For switching fabrics based on shared media, capacities up to 10 Gbit/s have been reported, and for shared memories, up to 5 Gbit/s. The semiconductor devices in the inputs and outputs basically set the limits of the fabrics, and the bit-rate handling capacities of such devices is still increasing.

In *space-division switch fabrics,* a number of paths are offered between the input and the output ports. Any of these paths can transmit cells between the input and output ports. Cells can travel on different paths at the same time. Thus the capacity of a space-division switch can be estimated as the average capacity of each path (or link) multiplied by the average number of paths that can carry traffic simultaneously. Thus these fabrics offer a high capacity that is limited only by such mechanical constraints as device pinout, connector restrictions, and synchronization restrictions.

Basically, cells can be routed through a network according to one of two methods: label routing or self routing. In *label routing,* a path for cells to traverse a network is set up in advance, i.e., at call setup. This creates a connection between an input port and an output port. This method is also called *preset path routing.* A prerequisite for this method is that the required bandwidth is known and that each link in the path is given sufficient bandwidth for the duration of the communication at call setup. In addition, each cell in the communication has to be routed over the selected path to a specified output port of each switching element using a virtual connection label or virtual channel identifier (VCI). Each cell has such a virtual channel identifier (VCI) that identifies the virtual channel to which the cell belongs. The name *label routing* comes from the VCI label.

As stated earlier, the asynchronous transfer mode (ATM) standard specifies the use of virtual paths and channels and is thus strictly a connection-oriented method. This implies the use of label routing. However, it is possible to deviate from ATM within a switch. This may be done to optimize the switch. One such method is called *self routing*, where no connection is set up in advance. Here, each cell has a *routing tag* that identifies the desired output port for that particular cell. This routing tag is analyzed in each switching element, and the identity of the destination switch output port is established. Through the use of one of several possible routing algorithms, the appropriate route for that cell is found, and the cell is guided through the network. Typically, the same table lookup mechanism is used for the routing tagging as for the VCI.

The two routing methods mentioned have their pros and cons. In the case of the method with a preselected path, the cons are that a set of suitable links has to be found, the switch elements in the path have to be marked, and the bandwidth of the path has to be managed with respect to required bandwidth and how to fulfill that request. The path also has to be disconnected and released at the end of the call. The bandwidth management in particular puts strains on the system in view of the variety of service requirements (voice, data, image, video, and mixes of these). The handling system also has to be fast in order to avoid traffic delays. On the pro side, we find that label routing is a better approach for multicast operation, i.e., when the same message is to be sent to several terminals. It is possible to combine self routing and label routing within the same switch and to use the first for point-to-point traffic and the second for multicast traffic.

A drawback with the self-routing method is that—because individual cells may take different routes—cells may arrive at the output port out of sequence. One way of handling this problem is to time stamp all cells when they arrive at the switch. At the output of the switch, the time stamps of the cells are watched, and the cells are buffered. A maximum delay time through the switch is established, and the cells are released from the output buffer at that maximum delay time. This ensures that the cells leave the switch in the same sequence as they arrived at it. Cells that encounter a delay beyond the established maximum delay cause trouble and should be discarded.

Several different network architectures for space-division switches exist. A popular network is the *banyan switching network* shown in Fig. 10.14. Typically, it consists of a number of switching elements with two inputs and two output ports, i.e., a 2 × 2 element or a degree

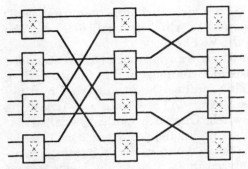

Figure 10.14 Banyan switching network.

two ($d = 2$) switching element. This size of a switching element is easy to make in both electrical and optical configurations. A banyan network has an order of complexity of $N \log N$, which is much less than that for a crossbar or matrix switch with a complexity of N^2. Banyan networks are blocking; the degree of blocking depends on the distribution of the requests for output ports in the incoming traffic, and it increases rapidly with the size of the network.

The performance of a banyan network can be improved by adding stages beyond the initial two stages, as shown in Fig. 10.15. As a result, there will be more than one path between designated input and output ports. Accordingly, these networks are called *multipath networks*. The multiplicity of paths also means that some mechanism to select one of the possible paths has to be incorporated.

In addition to the augmenting of a banyan switching network, the performance of a banyan network can be improved in two other ways. One result is a so-called Beneš network of the type shown in Fig. 10.16. Beneš (1965) worked for AT&T Bell Laboratories and wrote a

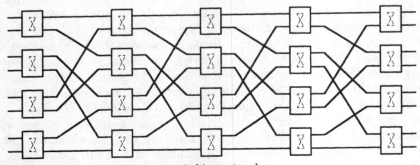

Figure 10.15 Augmented banyan switching network.

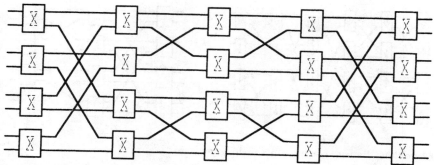

Figure 10.16 Beneš switching network.

book on telecommunications networks. A Beneš switching network basically consists of a multiplicity of banyan networks.

Another way involves the use of switching elements with more than two inputs and outputs. One arrangement of the latter type is the *delta network* shown in Fig. 10.17. Here, each input can reach each output through a single path. Delta networks are typically self routing and consist of two stages, as shown in the figure. As mentioned in the case of banyan networks, stages in addition to the first two can be added. If the degree of the switching elements is called d, a Delta network with N input (and output) ports requires $\log_d (N)$ stages with N/d switching elements per stage.

A combination of a banyan and a batcher network is shown in Fig. 10.18. A batcher network has the possibility of sorting the cells passing through it so that the cells at the output ports are arranged according to their routing tags. A banyan network that receives cells

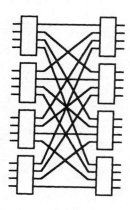

Figure 10.17 Delta switching network.

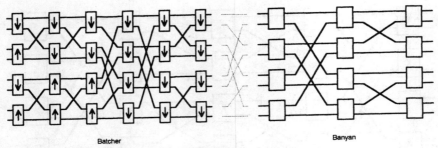

Batcher

Banyan

Figure 10.18 Batcher and banyan switching network.

arranged in this mode will offer nonblocking performance. Thus the combination of a sorting batcher network and a banyan network offers nonblocking performance unless multiple cells simultaneously request the same output port. The batcher network grows on the order of $N(\log N^2)$. As reported by Hickey (1990), a 256×256 batcher banyan network requires 36 sorting stages and 8 banyan switching stages. The implication is that these networks are more popular in the research communities than for commercial applications.

Clos (1953) gave his name to another type of network shown in Fig. 10.19. The type shown in the figure is a two-sided network. A Clos network also can be folded, as shown in Fig. 10.20.

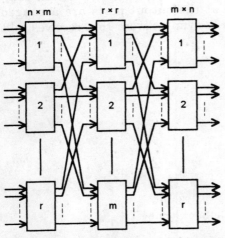

Figure 10.19 Two-sided Clos switching network.

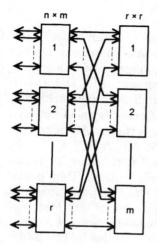

Figure 10.20   Folded Clos switching network.

Another way of increasing the performance of switching networks is to apply two or more networks in parallel. Actually, switch planes rather than switch networks are connected in parallel. Figure 10.21 shows an example of four switch planes arranged in parallel. Plane-selection mechanisms at the input and output sides distribute the traffic among the four planes.

10.4 Commercial ATM Switches

The ATM concept was originally intended for use in public broadband networks, in particular broadband integrated services digital networks (B-ISDNs). However, practical commercial applications started in private LAN interconnection networks.

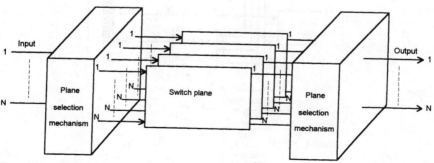

Figure 10.21   Switch planes in parallel.

10.4.1 The ASX-100 local switch

An example of an ATM switch designed for use to interconnect LAN networks and for other switching in small networks in general is Fore Systems' ASX-100. Figure 10.22 shows a block diagram of this switch. It consists of two main parts, the control processor and the switch board. ATM traffic enters and leaves the switch over network modules, of which there are a maximum of four. Different types of network modules are available for different kinds of interfaces. All offer full-duplex interfaces. A typical module has four 155-Mbit/s SONET OC-3c ports. Other varieties offer four 45-Mbit/s DS-3 ports, four 34-Mbit/s E-3 ports or one to six 100-Mbit/s multimode optical fiber ports. Modules for 622-Mbit/s full-duplex SONET OC-12c ports are also available. Each network module can support a total of 622 Mbit/s of full duplex traffic. Table 10.1 lists the possible combinations of ports.

The network modules are connected to queue modules by two 32-bit paths operating at 20 MHz, one for each direction. The queue mod-

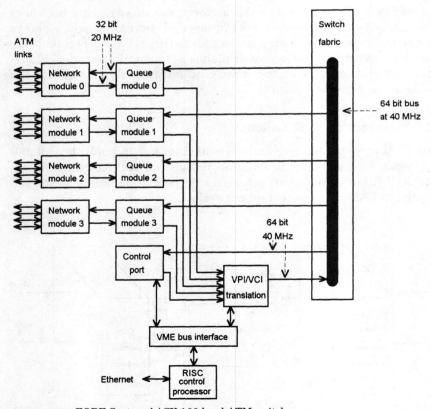

Figure 10.22   FORE Systems' ASX-100 local ATM switch.

TABLE 10.1   Number of ASX-100 Switch Ports by Network Module Type

45-Mbit/s DS-3 port	100-Mbit/s port	155-Mbit/s SONET OC-3c port	622-Mbit/s SONET OC12c port
16			
	24		
		16	
12	4/6		4
12		4	
12			1
8	4/6	4	
8	4/6		1
8		4	1
8			2
4	4/6	4	1
4		8	1
4	8/12	4	
4	8/12		1
4		8	1
4	4/6		2
4		4	2
4			3
	4/6		3
		4	3
			4

ules contain output buffers, flexible up to 4 Mbytes per port. They are used to release temporary congestion. Output buffers are also available at each network module port. From the queue modules, incoming traffic is fed to a VPI/VCI translator. It looks at the virtual path and channel identifiers of each cell, routes the cell accordingly, and translates and attaches the proper routing information to the cell for use on exit from the switch. In order to handle the maximum traffic from the four sets of network and queue modules, the single link from the VPI/VCI translator to the main switch fabric consists of a parallel 64-bit bus operating at 40 MHz. Similarly, the ATM switch fabric uses a 64-bit-wide time-multiplexed bus operating at 40 MHz. This means a throughput capacity of 2.5 Gbit/s. From the main switch fabric, traffic is directed to the proper queue module, on to the network module, and from there to the proper ATM output link. Multicast and broadcast traffic is sent to a multiple of output ports over one or several network/queue modules as required. Cells are copied at the output port level rather than at the switching fabric.

The ATM switch is controlled by an off-the-shelf RISC processor running a UNIX operating system. It interfaces with the VPI/VCI translator and a control port by way of a VME bus interface. The position of the control port in the switch architecture corresponds to and is similar to that of the queue modules. The control port funnels VPIs and VCIs between the main switch fabric, the VPI/VCI translator, and the control processor. The mapping tables used to translate VC identifiers and to route cells are contained in the switch board.

The overall capacity of the ASX-100 switch is limited by the capacity of the time division switch, in particular the 2.5-Gbit/s backbone bus. As a consequence, the switch is limited to a maximum of 24 ports, each with a bandwidth 155 Mbit/s, a maximum of six ports with a bandwidth of 622 Mbit/s, or a combination of these, as listed in Table 10.1.

FORE Systems is developing a 10 Gbit/s, nonblocking time-space-time switch.

10.4.2 Alcatel broadband switches

Alcatel-Alsthom is the largest worldwide manufacturer of telecommunications equipment (after having merged with the telecommunications part of ITT). The company's philosophy in ATM switching is to design modular units that can be combined in a range from small access and enterprise switches to large public switches. In addition, its policy—and that of several of its competitors—is to make it possible to substitute advanced future versions of the modules without causing premature obsolescence of already installed equipment.

Alcatel's basic ideal switching component is a module with 128 input ports and 128 output ports, as shown in Fig. 10.23. This module is initially implemented on a printed board assembly (PBA) and will later consist of a very large scale integrated circuit (VLSI), as discussed in Sec. 12.2.3. The external links of the modules are contemplated to be optical links operating at 622 Mbit/s. The economy improves with higher bit rates on these links. Within the module, switching is conducted on paths that operate at 155 Mbit/s. This bit rate is suited to the major bit rate of customer access lines at the current state of technology and applications.

In absence of the semiconductor technologies required for 128 × 128 modules, Alcatel is currently using modules with 64 × 64 input and output ports. Figure 10.24 shows this arrangement. Based on the modules described above, large switches with up to 16,000 ports can be assembled.

Most Alcatel switches use the multipath self-routing principle. This permits the switching module to act as a single-stage switching element even though it might consist of several individual switching elements. Thus the switching module can be arranged in different ways using integrated switching elements (ISEs). A 128 × 128 module can be arranged by eight 32 × 32 size switching elements. However, the technology is not ready for this size of ISEs. The first version of ISEs has a size of 16 × 16 ports (ISE 16). A switching module board with

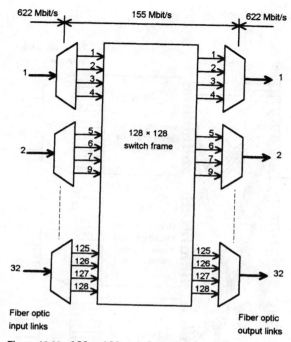

Figure 10.23 128×128 switch module.

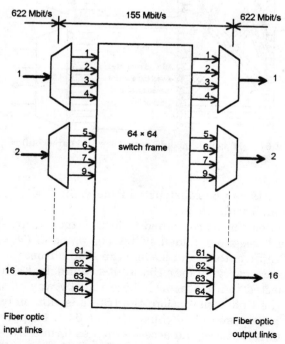

Figure 10.24 64×64 switch module.

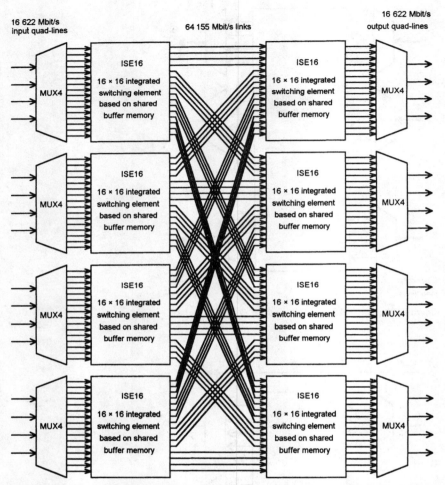

16 622 Mbit/s
input quad-lines

64 155 Mbit/s links

16 622 Mbit/s
output quad-lines

Figure 10.25 Two-stage Alcatel 64×64 switch module with 16×16 shared buffer memories.

64 input and 64 output ports can be constructed from eight ISE 16s. Figure 10.25 shows such an arrangement.

Several switching modules can be combined to form larger switches. Figure 10.26 shows a three-stage folded switch composed of 64×64 port switch frames. Traffic enters (and leaves) the switch from the left on 622-Mbit/s links, each carrying four 155-Mbit/s channels. Each of up to 32 links is connected to a traffic switching unit by way of a link termination. From the link termination, the traffic is randomly distributed to a set of access switches consisting of 64×64 port switch frames. The link terminations and access switches form a traffic-switching unit (TSU). There can be up to 128 TSUs.

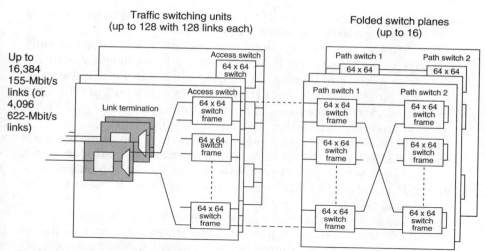

Figure 10.26 Alcatel 1000-series group switching network.

The access switches are connected to path switches on a folded switch plane consisting of up to two stages of 64×64 port switch frames called *path switches 1* and *2* (PS1 and PS2). The incoming traffic is randomly distributed over the PS1 switches and also over the PS2 switches. At the end of the PS2 switches, the traffic is directed "backwards" toward the proper port on the "left side" of the network. While having been randomly distributed toward the PS2 switches, the traffic is routed over designated links according to the routing tags in the cells toward the designated output port. The current design can have up to 16 folded switch planes.

The number of internal links is designed to account for twice the maximum possible traffic, thus offering underloaded links typically carrying 0.4 erlang each when the external links carry 0.8 erlang.

Rather than using the asynchronous transfer mode (ATM) internally, Alcatel uses its proprietary *multislot transfer mode* using *multislot cells* (MSCs). The ATM cells of 53 bytes can be said to occupy a 53-byte-wide slot on the transfer path. MSC uses smaller internal slots, which allows for smaller buffers. Multislot transfer mode allows for an optimized tradeoff between speed, delay, and parallelism. Besides ATM, the use of MSCs internally facilitates the adaptation of different cell sizes and the support of different traffic types at the edge of the switching network.

Figure 12.4 in Sec. 12.2.3 shows a block diagram of the integrated switching element (ISE). The ISE has 16 serial input ports, each with a bit rate of 155 Mbit/s. These ports convert the serial input to parallel signals with a width of 68 bits. These signals are placed on a 68-bit time-division multiplex (TDM) bus, which, in turn, transfers the signals to a shared buffer memory (SBM).

The SBM stores the MSC slots until they are ready to be transmitted over the output ports. The SBM is a dual-port memory with 256 slots with a width of 64 bits. When the time comes for the transmission of a particular MSC, it is transferred over the output 68-bit bus to the designated output port. This port converts the internal 64-bit-wide bit stream to a 155-Mbit/s serial bit stream.

In addition to the 16 input and output ports, a maintenance input port is connected to the 68-bit parallel bus on the input side, and a maintenance output port is connected to the bus on the output side. The ISE maintenance facilities are responsible for keeping the ISE operational.

Also connected to the 68-bit parallel input bus is the *input control*. It controls the information flow from the input ports as well as the maintenance port to the shared buffer memory (SBM) and the correct storage of 64-bit multislot cells (MSCs) in that memory. The input control also decodes slot control data, conducts error checks, and transfers the self-routing tag (SRT) to the routing logic.

The *routing logic* accepts the routing data from the input control and gives orders for the routing of the MSCs to the proper output port of the ISE. The routing information is interpreted differently depending on the actual ISE stage and the initial routing mode parameters. Basically, the traffic is distributed statistically in the first half of the switching network and is routed to the proper output port in the second half of the network.

The *output control* executes the orders of the routing logic and directs the MSCs to the proper output port. Finally, the output port converts the 68-bit parallel information back into a serial 155-Mbit/s data stream. The controller interface handles the interface between the ISE and external units, in particular the maintenance and test routines (MTR) and the on-board controller (OBC).

The links and the interface between users and Alcatel switches are shown in Fig. 10.27. Eight links with an access rate of 155 Mbit/s or two links with a rate of 622 Mbit/s connect users to a link termination that contains a semiconductor interface conversion chip. It converts the 1×622 Mbit/s or 4×155 Mbit/s serial bit streams coming from the users to an 8-bit parallel bus of the external protocol (or transmission) termination (ETT) unit.

The ETT handles all processing related to the external transmission system. It performs frame synchronization, scrambling and descrambling and/or coding and decoding of line signals, transmission overhead processing, monitoring, and policing when required. In the case of ATM, the virtual channel or path identifier is converted into a

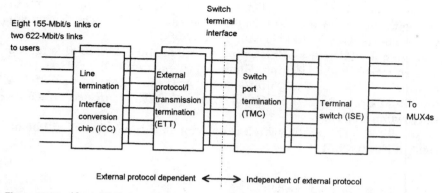

Figure 10.27   Alcatel link termination and switch/terminal interface.

generic reference number for the switch terminal interface. Some functions of the ATM layer such as cell delineation and header error checking and correction according to CCITT/ITU-T recommendations are also performed. The ETT is part of the switch module.

The ETT interfaces with the *switch port termination,* which contains the *transfer-mode converter* (TMC), a two-part proprietary chip consisting of TMC_{IN} and TMC_{OUT}. The ETT provides the TMC_{IN} with ATM cells and receives such cells from the TMC_{OUT} over 8-bit-wide parallel buses. The TMC_{IN} assembles the multislot cells (MSCs) by adding the self-routing tag and inserting the time stamp for resequencing. The TMC_{OUT} resequences the received multislot cells, converts the MSCs back to ATM cells, and provides queuing to the outgoing link. On the other side, the switch-port termination, as the name implies, acts as a termination of the switch modules.

Other link-termination functions include correct termination of the physical link, such as optical/electric conversion, signal regeneration, bit-clock recovery, frame synchronization, handling of overhead information, and physical looping.

Alcatel has produced VLSI chips for the ICCs, ETTs, TMC_{IN}s, and TMC_{OUT}s, which are described in Sec. 12.2.3.

10.4.3   AT&T ATM switches

AT&T Network Systems is offering and developing several different ATM switches within its Service Net-2000 family. The AT&T BNS-2000 broadband switch uses ATM cell technology on a shared bus with a throughput of 200 Mbit/s. The switch is basically designed to handle switched multimegabit data service (SMDS), but it can handle both connectionless and connection-oriented calls.

The AT&T GCNS-2000 ATM cell switch has a throughput of up to 20 Gbit/s based on a cell-relay/shared-memory fabric. The European version of the GCNS-2000 is called AXC-2000.

A version of AT&T's 5ESS switches called the 5ESS-2000 is a step toward broadband switching. However, the switch still switches DS-0 channels (56,000 or 64,000 bit/s each) with the potential to switch a bundle of up to 24 D-0 channels in parallel. This can accommodate narrowband video.

An AT&T ATM public switch is on trial at U.S. West and will be on trial in Europe in 1994.

10.4.4 Ericsson's pipe switch

Ericsson has developed hardware for a general-purpose ATM switch called the *ATM pipe switch,* which is the company's third generation of ATM switch hardware. The switch has a modular structure that allows for different sizes and applications. The switching functions are separated from the broadband applications so that the switching capability can be improved independent of the application. Hardware-dependent and hardware-independent software are separated by well-defined interfaces to allow for upgrading of software and hardware independently.

Figure 10.28 shows the layout of a typical Ericsson pipe switch. Users are connected to an access device (AD) that contains a switch port (SP). Different access devices are available depending of the application. The application-dependent part of the access device is separated from the rest of the switch by the cell bearer interface (CBI). The switch port (SP) handles the following functions: adaptation between the access device and the switch core, virtual path identifier/virtual channel identifier (VPI/VCI) translation, adding of routing information, and discarding of cells with an invalid VPI/VCI. Switch ports for bit rates of 155 and 622 Mbit/s and 2.4 Gbit/s are currently available. They support virtual circuit (VC) switching and virtual path (VP) switching simultaneously.

The switch ports are connected to switch-core planes through an ATM core interface. The switch core (SC) is a space switch that supports both point-to-point and point-to-multipoint connections. As indicated in Fig. 10.28, the switch-core plane can be duplicated to increase the availability of the switch.

The major space-switching elements in the Ericsson pipe switch are the ATM input circuit (AIC) and the ATM output circuit (AOC). The AIC accepts the incoming cell flow, typically from 32 links, aligns the cells, and converts the individual flows on the links to a single "pipe flow" of cells. Besides 32 links with a bit rate of 155 Mbit/s, the input links to an AIC can consist of eight 622-Mbit/s links or two 2.5-Gbit/s links.

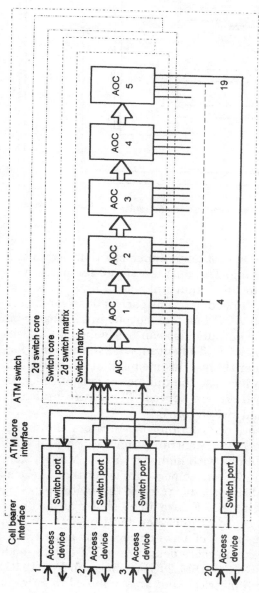

Figure 10.28 Ericsson single-board ATM pipe switch.

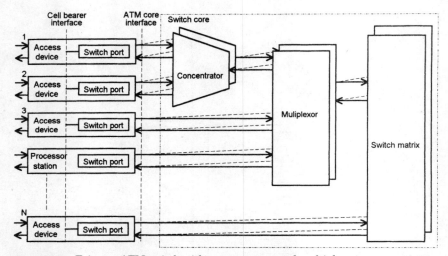

Figure 10.29   Ericsson ATM switch with concentrators and multiplexors.

The ATM output circuit (AOC) contains a memory that acts like a buffer, and it performs switching. The cells are stored temporarily in the buffer and are read out to the output links according to the VPI and VCI information in the cell headers. Thus each cell is switched to the correct output link. Each AOC has four output links with a bit rate of 155 or 622 Mbit/s. The outputs also can be combined to a single link with a bit rate of 2.5 Gbit/s. The outputs are in the same format as the inputs. Among other things, this means that several AOCs can be connected in series to increase the number of output links. In Fig. 10.28, 20 input/output links are served. The figure indicates duplicated switch matrices and switch cores.

An Ericsson ATM switch core can contain concentrators and multiplexors, as shown in Fig. 10.29. A concentrator combines several links with low bit rates into fewer links with higher bit rates. Multiplexors are used when the bit rates at the switch ports are lower than those at the switch matrix, and they bring the bit rates of the switch ports up to the bit rate of the switch matrix. Figure 10.29 shows an arrangement of a switch core with concentrators and multiplexors. Figure 10.30 shows a possible arrangement of a switch core with a multiplexor and a possible switch matrix in more detail. An ATM switch module with a throughput of 20 Gbit/s (ASM 20G) is shown in Fig. 10.31. It consists of four 4×4 switch matrices that are combined to form an 8×8 switch matrix. Each of the links can handle traffic of 2.5 Gbit/s. The incoming traffic on the 2.5-Gbit/s links from the left is split between two 4×4 matrixes horizontally. Similarly, the outputs from

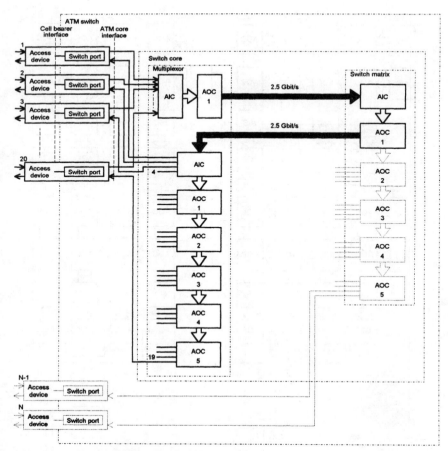

Figure 10.30 Ericsson ATM switch with multiplexor.

two 4×4 matrices are merged vertically. ASM 20G units are combined to form switches with capacities beyond 20 Gbit/s. Four ASM 20G units create a unit with a throughput of 40 Gbit/s in a quadratic expansion in a manner similar to the expansion in the ASM 20G module. An ASM 40G can handle 16 links at 2.5 Gbit/s or, with multiplexors, 256 links at 155 Mbit/s. Similarly, four ASM 40G units can be expanded to handle a throughput of 80 Gbit/s in an ASM 80G unit.

10.4.5 Fujitsu FETEX-150

The Fujitsu electronic telecommunication exchange FETEX-150 has a modular architecture. It is designed to evolve from a conventional narrowband signal path system platform to platforms with broadband

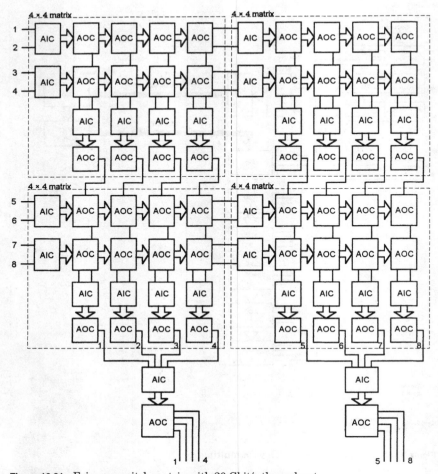

Figure 10.31 Ericsson switch matrix with 20 Gbit/s throughput.

signal path subsystems. Figure 10.32 depicts the possible arrangements. Fujitsu is of the opinion that the conventional synchronous
transfer mode (STM) is preferable to asynchronous transfer mode
(ATM) for continuous traffic such as the distribution of television programs. Thus the FETEX-150 switch may be equipped with an STM
platform. The switch also will contain an ATM platform for bursty
traffic. Fujitsu visualizes that ATM will dominate in the far future.
Both connection-oriented (switching signals from 64 kbit/s to 155
Mbit/s) and connectionless ATM service will be offered. Subscriber
interfaces will be offered at 155 and 622 Mbit/s. To save on optical
links, the FETEX-150 system offers broadband remote switching

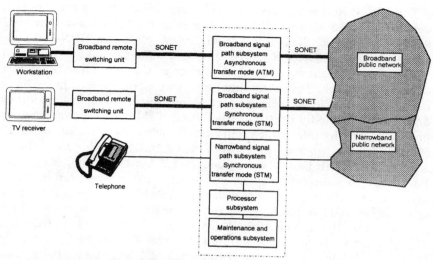

Figure 10.32 Fujitsu FETEX-150 switch configuration.

units to be located close to or even on the customer premises. Network interfaces will be of the SONET type.

10.4.6   NEC NEAX 61E ATM service node

The ATM switch offered by NEC and NEC America, Inc., is similar to the FETEX-150 offered by Fujitsu in the following aspects: both are modular, both offer an evolution from narrowband to broadband switching, and both offer both asynchronous (ATM) and synchronous transfer mode (STM).

10.4.7   Northern Telecom Magellan family

Over the course of 1993 and 1994 Northern Telecom announced a portfolio of switching products under the brand name "Magellan." These products include the Magellan Access Switch, Magellan Passport enterprise network switch, Magellan DPN-100 switching system, Magellan Gateway ATM carrier access switch, and Magellan Concorde ATM carrier backbone switch.

The Magellan Access Switch is designed to consolidate traffic in remote business offices. It supports mixed-protocol switching of both traditional data applications such as SNA, ASync, Bisync, and the LAN bridging and routing applications. The Access Swtich supports speeds up to 256 kbit/s.

The DPN-100 is a frame-based wide area data network system. It is designed for SNA internetworks and as a multiservice traditional data (X.25, Async, SDLC, and frame relay) platform. DPN-100 supports trunking and access speeds of up to 2.0 Mbit/s.

Figure 10.33 shows a block diagram of the Magellan Passport enterprise switch. Different function processors are available for a range of applications, including trunking, voice, video, and different data and multimedia protocols, including ATM. One type of function processor can handle frame relay traffic with bit rates of up to 45 Mbit/s. Future function processors will handle OC-3c (155 Mbit/s) ATM traffic. The backplane speed is 1.6 Gbit/s. Thus, Magellan Passport can support up to 10 OC-3c ports. A queue manager can arrange for buffering of low priority traffic while high priority traffic is switched. It also manages delays in constant bit rate traffic such as voice and video. The Magellan Gateway broadband multimedia ATM switch is designed for public networks. The current version of Magellan Gateway provides a throughput of 1.2 Gbit/s and supports up to 24 DS-3 ports or a combination of DS-1, DS-3, and OC-3c ports. Figure 10.34 shows a block diagram of the switch. External interface modules (EIMs) supply user, network, and service adaptation interfaces at 155 Mbit/s each. Gateway offers eight slots for EIM cards, which can interface with DS-3, OC-3, and DS-1 lines. The switch is not constrained to the allocations shown in Fig. 10.34 and the eight EIMs can be used for any type of interface. From the

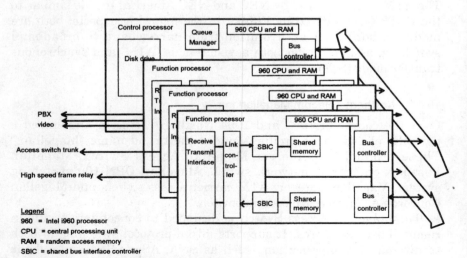

Figure 10.33 Northern Telecom Magellan Passport switch.

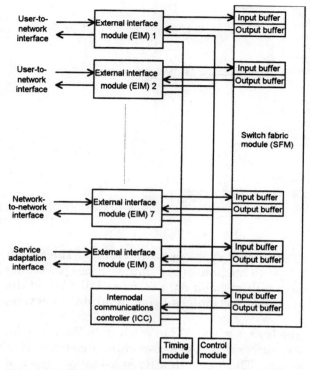

Figure 10.34 Northern Telecom Magellan Gateway switch.

EIM, the ATM traffic is directed to input buffers in the switch fabric. The fabric is an 8×8 switch. After switching, the traffic leaves the switch fabric through an output buffer. Redundant control modules run the node control software and send messages to other switches through the internodal communications controller (ICC), injecting control information into the normal traffic streams.

The Magellan Concorde is a high performance ATM switching system with common memory buffer and 40 Gbit/s capacity. The switch is scalable down to 10 Gbit/s. The Concorde supports ATM user access at DS-3, OC-3c, OC-12, and OC -48 rates.

10.4.8 Siemens broadband switching systems

Figure 10.35 shows the features available on Siemens' EWSM cell switching system. Customer interface bit rates range from 64 kbit/s to

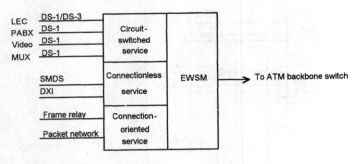

DXI = data exchange interface

SMDS = switched multi-megabit data service

Figure 10.35 Siemens, EWSM cell switching system.

45 Mbit/s, and network interfaces range up to 155 Mbit/s. The switch is mainly designed for low- to medium-speed data services, including frame relay and switched multimegabit data service (SMDS) (and the corresponding European connectionless broadband data service, CBDS).

Figure 10.36 shows the layout of the ATM switch from Siemens. An ATM switching network performs the ATM switching functions. It is self routing and nonblocking. This network is accessed by two types of ATM cards, A and B. Type A is used for subscriber line and type B for trunk traffic. Both types of cards interface with a multiplexor and the ATM switching network at 175 Mbit/s. All B-ISDN specific functions (routing control, call setup and release, feature control, call charging, and statistics) are performed by software resident in the group processor(s). Operation, administration, and maintenance (OAM) features are included, and the switch can be connected to a telecommunications management network (TMN).

10.5 Cross-Connect Systems

Digital cross-connect systems for interconnecting and rearranging carrier channels such as T1 and T3 have been available for many years. Similar systems for use on fiberoptic links are now available. They can handle SONET and/or SDH links. Most such cross-connect systems are operated manually; i.e., an operator rearranges links manually. Semiautomatic and automatic systems are becoming available. The main difference between an automatic cross-connect system and a switch is that the first is handled by the service provider, while a switch can be handled by the user through dial-up.

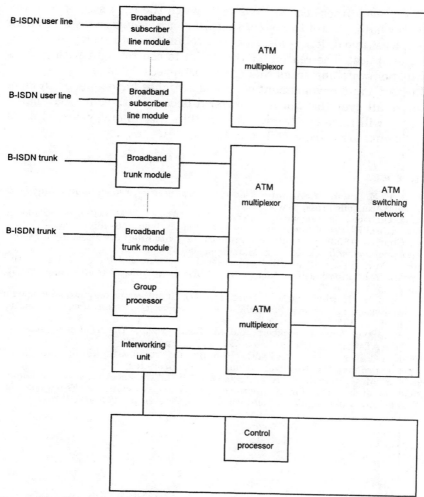

Figure 10.36   Siemens' ATM switch.

10.6   Optical Switches

Optical fibers are used extensively as broadband transmission links in telecommunications applications. With the exception of such links, however, all other transmission links carry electrical signals rather than light signals. All switching functions are conducted on electrical signals. This means that electrical signals have to be converted to light signals at the beginning of each optical link, and these must be converted back to electrical signals at the end of each link. It seems logical to try to avoid these conversions and switch the light signals directly. This is what is meant by *optical switching* or *photonic switching*. There are at least two different means of controlling an optical switch: electrical and optical. The reason for the desire to control light with light is the higher switching speed that can be achieved.

Research and development of optical switches have gone on in laboratories all over the world for several decades. It seems reasonable that we will have optical switches within a couple of years. Optical components are discussed in Sec. 12.4.

References

Beneš, V. E. (1965), *Mathematical Theory of Connecting Networks and Telephone Traffic,* Academic Press, New York.

Boettle, Dietrich, and Michel A. Henrion (1990), "Alcatel ATM switch fabric and its properties," *Electrical Communication,* vol. 64, no. 2/3, pp. 156–165.

Clos, Charles (1953), "A study of non-blocking switching networks," *Bell System Technical Journal,* vol. 32, no. 2, March, pp. 406–424.

Henrion, Michel A., Gert J. Eilenberger, Guido H. Petit, and Pierre H. Parmentier (1993), "A multipath self-routing switch," *IEEE Communications Magazine,* vol. 31, no. 4, April, pp. 46–52.

Karol, M. J., M. Hluchyi, and S. P. Morgan (1987), "Input versus output queueing on a space-division packet switch," *IEEE Transactions on Communications,* vol. 35, no. 12, pp. 1347–1356.

Larsson, Mikael, Martin Ljungberg, and Jan Rooth (1993), "The ATM switch concept and the ATM pipe switch," *Ericsson Review,* no. 1, pp. 12–20.

Newman, Peter (1992), "ATM technology for corporate networks," *IEEE Communications Magazine,* vol. 30, no. 4, April, pp. 90–101.

Yeh, Y. S., M. G. Hluhyi, and A. S. Acampora (1987), "The knockout switch: A simple modular architecture for high-performance packet switching," *IEEE Journal on Selected Areas in Communications,* vol. 5, no. 8, October, pp. 1274–1284.

Internetworking

11.1 Introduction

A company often finds itself with many different local area networks (LANs) operating within the organization. Not only are the LANs separate, they often also come from different vendors. Among reasons for this fact are that different departments acquired LANs from different vendors at different times, different types of LANs came with acquisitions and mergers, and the companies often actually want to keep the LANs separate between different departments, such as accounting, marketing, R&D, and corporate planning.

In time, demand for communications between terminals located on different LANs arises. This includes demand for communications capabilities between LANs located near each other, as well as demand between facilities at great distances from each other and even on different continents.

In talking of interconnecting LANs, three different levels are commonly used: local networking, metropolitan area networks (MANs), and wide area networks (WANs). They can be differentiated by the maximum distance between terminals connected to the networks. This distance can be a few kilometers for local networking systems and up to 80 km for MANs, while a WAN can span the entire earth.

11.2 Interconnection of Local Area Networks (LANs)

11.2.1 Introduction

Different means of interconnecting LANs exist. Among them are bridges, routers, and gateways. A *bridge* interconnects LANs of the

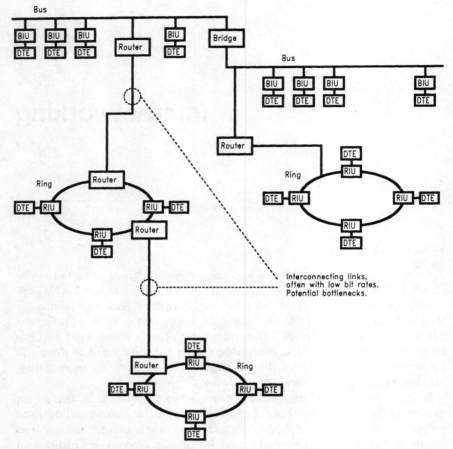

Figure 11.1 Interconnection of LANs.

same type using the same protocols. As an example, the two bus-structure LANs at the top of Fig. 11.1 are supposed to use the same protocols and are connected by a bridge.

11.2.2 Bridges

A *bridge* operates on the first two layers (physical and medium access) of the International Organization for Standards Open Systems Interconnection (ISO-OSI) model, as shown in Fig. 11.2. The bridge reads all frames transmitted on the LAN to the left and retransmits those addressed to the LAN on the right on that LAN. Similarly, data are transmitted in the opposite direction. Originally, bridges required the protocols on the two sides of the bridge to be

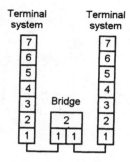

Figure 11.2 Protocol interface through a bridge.

identical. Recent bridges can interconnect LANs with slightly different protocols as long as their medium access control (MAC) protocols have very similar MAC frame formats (such as CSMA/CD and token bus). In this case, the bridge uses the MAC protocol of the addressed LAN for transmission to that LAN. If many LANs are interconnected by bridges, each bridge must contain enough intelligence to decide on the routing of the payload information.

11.2.3 Routers

A *router* can interconnect LANs of different types, including those using different protocols. Routers are shown between the ring networks and between those and the bus networks in Fig. 11.1. Besides layers 1 and 2 of the ISO-OSI model, a router involves layer 3, as shown in Fig. 11.3. If the two networks are close together, a single router suffices. On long interconnection links, one half of a router is usually located at each end of the link. Bridges and routers are used when the LANs use compatible protocols from the OSI model.

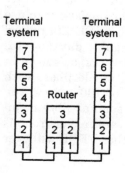

Figure 11.3 Protocol interface through a router.

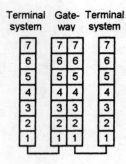

Terminal   Gate- Terminal
system   way   system

Figure 11.4 Protocol interface through a gateway.

11.2.4 Gateways

In the case of incompatible proprietary protocols (such as IBM's SNA), gateways are required. A *gateway* performs protocol conversions on the application level, utilizing all the lower layers, as shown in Fig. 11.4.

Typically, bridges and routers cannot extend the overall transmission distances much beyond those of the individual connected LANs. The maximum transmission rate between LANs that are interconnected by bridges or routers is limited to the rate of the lowest-rated LAN. Often the rate is further limited by the transmission capability of the link between the LANs, as in the example shown in Fig. 11.1.

Such a link can theoretically consist of a simple telephone line with a bit rate of 300 bit/s. When we compare this to the bit rate of 10 Mbit/s within a LAN, problems become obvious. Special solutions are required not just because a LAN *can* transmit at a high rate, but mainly because the users demand a much higher rate.

11.2.5 FDDI follow-on LAN (FFOL)

Realizing the need for a high-speed backbone network to connect FDDI and FDDI-II networks, the FDDI standards committee is working on a new standard, tentatively called *FDDI follow-on LAN* (FFOL) (Jain, 1993). Besides FDDI and FDDI-II, the purpose is to use it for interconnecting Ethernet, token ring, and other IEEE Standard 802-type networks, as well as switched multimegabit data service (SMDS) networks, broadband ISDN, and ATM networks, as shown in Fig. 11.5. FFOL is expected to use available multimode fibers with speeds of 100 Mbit/s to 2.5 Gbit/s over distances of 100 m to 2 km. FFOL will compete with other gigabit standards such as fibre channel, high-performance parallel interface (HIPPI), and asynchronous transfer mode (ATM). It is thus questionable whether FFOL will ever make it to the market.

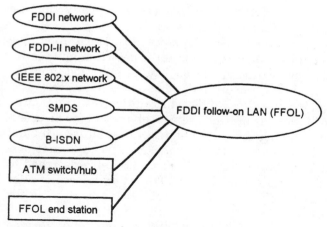

Figure 11.5   FDDI follow-on LAN backbone.

11.3   Metropolitan and Wide Area Networks (MANs and WANs)

11.3.1   Introduction

Two technologies specified for MANs, FDDI and DQDB, use a shared transmission medium, while other forms of MANs and all WANs use packet switching. Besides this difference between MANs and WANs, there is no reason for discussing them separately.

Private digital lines used to interconnect LANs can be had at practically any desired transmission rate, but at a very high cost. In order to reduce the cost of such interconnecting links, a demand for dial-up links with high bit rates is emerging. Among solutions are *frame relay, switched multimegabit data service* (SMDS), narrowband ISDN, asynchronous transfer mode (ATM), and broadband ISDN. Of these, frame relay and SMDS can be considered a stop-gap offering in anticipation of faster services such as ATM and B-ISDN. While fiber distributed data interface (FDDI) is designed as a LAN service, some providers propose to offer it as a WAN service.

11.3.2   Frame relay

Traditional packet switching, e.g., CCITT Recommendation X.25, was developed when long-distance transmission facilities had very bad quality. To compensate for that, the packet-switching protocols had error detection and correction built in. These features created significant overhead in the transmitted bit stream. With today's improved transmission lines, it is considered unnecessary to handle errors at the lower ISO-OSI layers. Rather, error detection and correction are referred to higher layers in the protocol stack.

Frame relay was developed as a consequence of this reasoning. The frame-relay overhead has been stripped of error detection and correction features. With less overhead, the payload is enlarged, and the ratio of payload to total bits transmitted is increased. In other words, the effective payload bit rate is expanded. In addition to stripping the overhead from error functions in frame relay, call control (such as setting up and releasing connections, etc.) has been moved from inband signaling (in X.25) to a separate logical connection. The format of a frame-relay packet is shown in Fig. 3.27. Frame relay is designed to operate at bit rates of up to about 2 Mbit/s. Thus it still does not offer broadband connectivity. Further, frame relay cannot be used for transmission of voice and video.

In the latest developments, international frame-relay service between continents is being considered.

11.3.3   Switched multimegabit data service

Bell Communications Research, Inc. (Bellcore), the research company owned by the seven regional Bell operating companies (RBOCs) in the United States, developed the concept and technology of *switched multimegabit data service* (SMDS). As the name implies, SMDS is designed for data, and it cannot easily be used for the transmission of voice or video. Most RBOCs are offering the service or are in the process of introducing it. The European Telecommunications Standards Institute (ESTI) has standardized a similar feature called *connectionless broadband data service* (CBDS).

The user-network interface (UNI) used by SMDS is based on the distributed-queue dual-bus (DQDB) protocol described in IEEE 802.6. Figure 3.27 shows a DQDB frame. The UNI provides access to a switched network using bit rates of 1.2 up to 34 Mbit/s. Future extension to 155 Mbit/s is planned. Currently, access can be provided with DS-1 (T1) and DS-3 (T3) channels, while optical OC channels will be used in the future. In the United States, users can subscribe to different access classes (ACs) that specify the maximum sustained bit rate. SMDS is actually a software package that operates over existing transmission facilities, such as DS-3 and the North American Synchronous Optical Network (SONET). Thus it cannot really offer bit rates higher than those offered by DS-3 and SONET.

An example of an SMDS network and the protocols used is shown in Fig. 11.6. The traditional protocol stack of end system A is shown to the left. An imaginary user is connected to a LAN, and the LAN is connected to a data service unit/channel service unit (DSU/CSU)

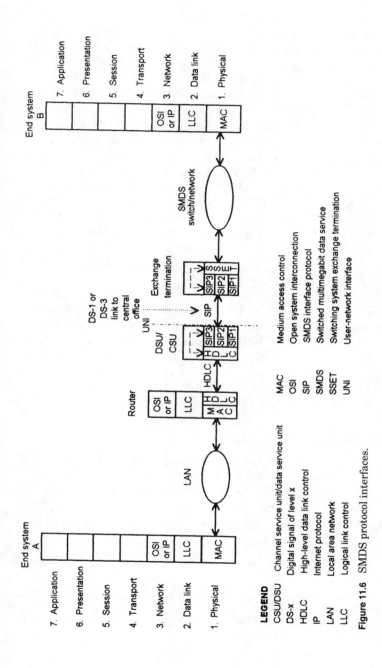

Figure 11.6 SMDS protocol interfaces.

LEGEND

CSU/DSU	Channel service unit/data service unit
DS-x	Digital signal of level x
HDLC	High-level data link control
IP	Internet protocol
LAN	Local area network
LLC	Logical link control
MAC	Medium access control
OSI	Open system interconnection
SIP	SMDS interface protocol
SMDS	Switched multimegabit data service
SSET	Switching system exchange termination
UNI	User-network interface

through a router. The router is required if the protocols used on each side of it are different. As usual, the accommodation of the different protocols is based on OSI layers 1 and 2 and performed at layer 3. The router is connected to the DSU/CSU by a link using the high-level data-link control (HDLC) protocol according to established conventions. The user-network interface (UNI) is found on the right side of the DSU/CSU. The type of link between the DSU/CSU and the termination of the SMDS switching system, the so-called exchange termination (ET), is determined by the expected traffic over that link. It can be a T1 system with a DS-1 channel, a T3 system with a DS-3 channel, or an optical system. An SMDS interface protocol (SIP) operates between the user access termination inside the CSU and an SMDS switch or switching network. The SIP protocol has three levels (which are different from the OSI layers) and is based on the DQDB protocol. Basically, SIP level 1 provides physical layer functions, level 2 handles error detection and framing, and level 3 contains addressing information, etc. An SIP level 2 protocol data unit is shown in Figure 3.27. To simplify Fig. 11.6, user B to the right is connected directly to an SMDS switch.

An SMDS switch can be an improved version of a conventional packet switch, one based on the DQDB standard (a so-called metropolitan area network switch), or an ATM switch. Any of the packet switches can be converted to handle voice and video over an SMDS network by adding output buffers that delay the packets until they can be released in their original order. Another approach is to time stamp each packet at the point of origin and to release them at a fixed time delay at the destination output.

11.3.4   Asynchronous Transfer Mode (ATM)

When terminals connected to LANs start to generate voice and video traffic, when the demand requires higher bit rates, and when such traffic becomes integrated with data traffic, frame relay and SMDS will become insufficient. Rather, the trend will turn to the use of ATM and broadband ISDN. These approaches are the major subject of this book, and they are discussed elsewhere.

References

Jain, Raj (1993), "FDDI: Current issues and future plans," *IEEE Communications Magazine,* vol. 31, no. 9, September, pp. 103–104.

Stallings, William (1993), *Advances in Local and Metropolitan Area Networks,* IEEE Computer Society Press, Los Alamitos, Calif.

Building Blocks

Chapter

12

Components

12.1 Introduction

Digital broadband technology is still in the initial phase. This means
that manufacturers are doing their own research and development
and that few components have been standardized. In other words, the
manufacturers use their own proprietary components. Future hardware for digital broadband equipment will to a high degree consist of
semiconductor devices.

12.2 Semiconductors

12.2.1 Introduction

In particular, such semiconductor devices as very large scale integrated (VLSI) circuits and application-specific integrated circuits (ASICs)
will be used in broadband equipment. Today, a semiconductor device
can consist of a single transistor or be a VLSI device with hundreds of
thousands of transistors. Researchers at IBM have put an entire
large computer on a single chip. Thus it may be possible eventually to
put an entire ATM switch on a single chip. Today, several chips are
required for switches, multiplexors, interface devices, etc. Some
announced and/or commercially available semiconductor devices are
described in the following paragraphs.

Broadband applications require logical devices, such as digital
switches, and memories that operate at bit rates of tens of gigabits
per second (Gbit/s). The technology constraints of such devices,
besides speed, include integration level and complexity, power consumption, availability, and cost.

The types of semiconductor devices being considered for broad-d applications include complementary metal-oxide semiconductors (CMOSs), bipolar semiconductors, and combined bipolar and CMOSs, so-called BICMOSs. Devices based on both silicon and gallium arsenide (GaAs) are considered. The time from pilot production in a laboratory of a new type of semiconductor to commercial availability can range from several years to tens of years. In other words, even though a new device is discussed in the literature, it may not be available for application in commercial products for a long time.

Semiconductors are based on the introduction of chemical impurities into the crystal lattice of the surface of the device. These impurities create p-type or n-type areas in the surface according to desired patterns. The use of different patterns and the layering of impurities result in different types of semiconductors.

The conventional steps in fabricating silicon integrated circuits typically include preparation of the silicon material, deposition of impurities (epitaxial growth) to create p-type or n-type silicon substrates, and surface preparation (passivation) through oxidation (to create SiO_2) or to form silicon nitride (Si_3N_4), as shown in Fig. 12.1. In following steps, the passivated wafer is coated with a photoresist, covered with a mask, exposed to radiation, developed, etched, and stripped of the photoresist. The masks are produced through photolithography and are used to create patterns in the photoresist on the wafer surface. Depending on whether positive or negative resists are used, negative or positive patterns are obtained in the photoresist. During the following etching process, nondesirable areas of the passivation layer are removed, leaving a desired pattern open to exposure by impurity ions. Finally, the photoresist is removed (stripped).

In a following step, a controlled number of impurities are entered into the unprotected areas of the wafer surface through one of two processes, diffusion or ion implantation. In the diffusion process, the wafers are exposed to impurities in a diffusion furnace. For ion implantation, the wafers are bombarded with impurity ions in a vacuum chamber.

Logical CMOS devices have higher switching speeds and less power dissipation compared with MOS devices. Emitter-coupled logic (ECL) gates offer the highest clock rate.

Most semiconductor memories are either bipolar or metal-oxide (MOS). Of these, bipolar memories offer higher speeds and signal levels than MOS devices. On the other hand, the power consumption of MOS devices, and in particular CMOS devices, is lower than that of bipolar devices.

The switching speed of a semiconductor device depends on the fea-

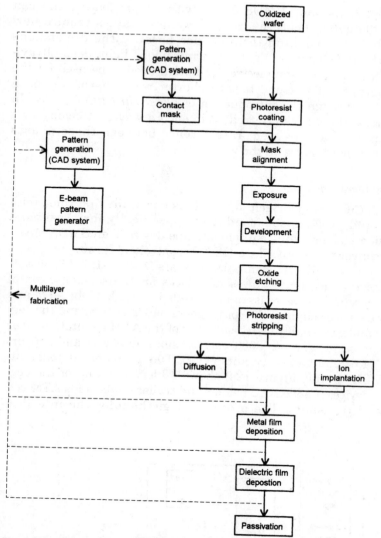

Figure 12.1 Semiconductor fabrication.

ture sizes (line width, spacing, channel length, and gate length), package density, and power consumption. The speed can be increased with smaller feature sizes, less density (number of gates), and more power. Current feature sizes for CMOS devices are about 0.8 µm, while sizes of 0.6 to 0.5 µm are expected in a few years. The minimum feature sizes for BICMOS are slightly higher at 1.2 and 0.8 to 0.6 µm, respectively. A 150-mm² CMOS device with 0.6-µm feature size can have 350,000 gates. Feature sizes below these values require new lithographic and masking technologies because the dimensions approach the wavelength of light. Thus ultraviolet light, deep-ultraviolet light, x-rays, and even electron radiation need to be used for the photolithography. The trend toward shorter wavelengths has been slowed down through the introduction of *phase-shift lithography*. In this approach, the masks are treated in such a way that beams with different phases are created. Interference between these phases results in sharper images on the wafer.

12.2.2   TranSwitch devices

TranSwitch Corporation (Shelton, Conn.) is one of the first companies to design and produce semiconductors specifically for broadband transmission and switching equipment. One device is an SMDS physical layer convergence procedure (PLCP) controller. Its function is to map cells into one of the transmission formats DS-1, E-1, DS-3, or E-3.

The company also makes a set of devices for the segmentation of packets into cells and the reassembly of cells into packets for ATM and SMDS. The devices support the convergence sublayer (CS) and the segmentation and reassembly (SAR) sublayer of the ATM adaptation layer (AAL). They segment up to 8000 packets simultaneously and support from 512 to 64,000 virtual circuits (VCs). The set manages constant-bit-rate (CBR) traffic. Figure 12.2 shows a block diagram of the segmentation device. It adds AAL header and trailer fields, adds ATM cell header, and transmits. The processor logic handles the interface

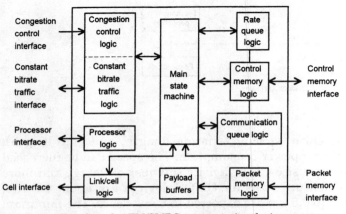

Figure 12.2   TranSwitch ATM/SMDS segmentation device.

between the local host/processor and the internal registers of the device. The output cell stream leaves the device over the cell interface. The control memory interface is used to access the control memory. Variable-bit-rate traffic packets are read into the device over the packet memory interface. Constant-bit-rate traffic enters the device over the CBR interface. The reassembly device sends congestion information to throttle the source cell rate over the congestion control interface.

Figure 12.3 shows a block diagram of the reassembly device. The internal registers of the device are accessed through the processor interface. Bus arbitration signals for access to the control memory and data-transfer signals enter the device over the control memory interface. The descriptor control logic maintains working registers and processing logic for the buffer descriptor manipulation. It maintains the various base address registers and generates the control memory address and data bus signals. It also contains a timer for aging packets. The communications queue logic contains all the pointers and associated pointer logic for the four communication queues in the control memory and the two-cell FIFOs in the packet memory. The pointers can be accessed by the host through the processor/host interface. The packet memory control logic interfaces with the packet memory. It generates all the control signals for bus arbitration and data transfer to the packet memory. It also generates the packet memory address and transfers data from the cell buffer to the packet memory. It can perform both synchronous and asynchronous data transfers. The main state machine controls six of the blocks, as shown in Fig. 12.3. It generates all the control signals and sequences that operate the device. The state machine is implemented as a state machine of 64 states.

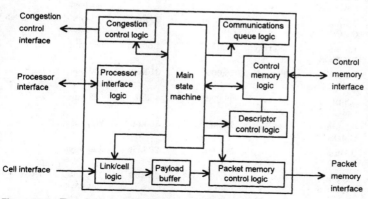

Figure 12.3 TranSwitch ATM/SMDS reassembly device.

Input cells are read into the reassembly device over the cell inter-face. The processor/host reads and writes the internal registers and states of the unit over the processor interface. The virtual circuit (VC) lookup table, the buffer descriptor table, and the various communica-tions queues in the control memory are accessed through the control memory interface. The packet memory interface transfers cells to the packet memory, where these cells are reassembled into packets.

12.2.3   Alcatel

Alcatel has designed several VLSI chips for the company's ATM switches, including the integrated switching element (ISE), the exter-nal transmission termination (ETT), the interface conversion chip (ICC), and the TMC_{IN} and TMC_{OUT}. Table 12.1 shows the characteris-tics of some of the Alcatel devices.

In the early 1990s, the technology allowed for the production of inte-grated switching modules (ISEs) with 16×16 ports implemented in 0.8-μm CMOS technology only. In a second phase, 32×32 port ISEs employing 0.5-μm technology are used. BICMOS technology is required for multiplexors operating at bit rates of 622 Mbit/s and for clock conver-sion and distribution at 622 MHz. Other devices, such as switch port ter-minations (TMCs) and external transmission terminations (ETTs), are less demanding and have been made in CMOS with 0.8-μm features.

Alcatel switching modules can be arranged in different ways using integrated switching elements (ISEs) designed and made as very

TABLE 12.1   Some Alcatel VLSI Broadband Telecommunications Chips

Chip	Technology	Design	Gates	Memory	Package	Power
ISE16	0.8-μm CMOS		55,000	25 kbit	128-pin flat pack	6 W @ 5 V
ISE32	0.5-μm CMOS		100,000	30 kbit	196-pin flat pack	6 W @ 3.3 V
ISE receive-transmit port	1.5-μm BICMOS	Gate array	11,000	1.5 kbit	132 pins	1.9 W
ISE central memory chip	1.5-μm CMOS	Cell based	16,000	31 kbit	132 pins	1.2 W
TMC_{IN}	0.8-μm CMOS		20,000	4 kbit	132 pins	2.3 W
TMC_{OUT}	0.8-μm CMOS		25,000	70 kbit	80 pins	2.8 W
ETT	0.8-μm CMOS		50,000	20 kbit	144 pins	5.5 W
Multiplexor	0.8-μm BICMOS		14,000 CMOS 1,500 bipolar	16 kbit	196-pin flat pack	3.1 W @ 3.3V

SOURCE: *Electrical Communication,* vol. 64, no. 2/3, pp. 169–171.

large scale integrated (VLSI) circuits. A 128×128 module can be arranged by eight 32×32 size switching elements. However, as stated earlier, the technology is not ready for this size of ISEs. The first version of ISEs has a port size of 16×16 (ISE 16). A switching module board with 64 input and 64 output ports can be constructed from eight ISE 16s.

The switching feature of the Alcatel integrated switching elements is based on a shared memory, as shown in Fig. 12.4. The unit has 16 serial input ports each with a bit rate of 155 Mbit/s. These ports convert the serial input to parallel signals with a width of 68 bits. These signals are placed on a 68-bit time-division multiplex (TDM) bus, which, in turn, transfers the signals to a shared-buffer memory (SBM).

12.3.4   Other manufacturers

Practically all manufacturers of broadband equipment are involved with design and—to a lesser extent—manufacture of relevant semiconductors. Many of them are based on application-specific integrated circuits (ASICs). Typically, a semiconductor fabrication house (fab) offers integrated circuit (IC) designs as a kind of semifinished product, to which a user can make modifications and additions to arrive at a finished IC. This reduces the costs through the efficient use of the fab's facilities. Once integrated circuits used in broadband equipment become standard, semiconductor houses are expected to start producing these standard devices.

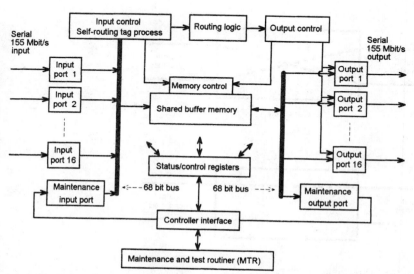

Figure 12.4   Alcatel integrated switching element (ISE).

12.3 Fibers

Optical fibers used to transmit optical signals in telecommunications are made of very pure silica. Dopants are added to control the index of refraction. They are typically 125 µm thick. The fiber construction consists of a doped silica core with a glass cladding. The cladding has a refraction index that is 0.3 to 1 percent lower than that of the core. This keeps the light inside the fiber. Using a thick core (>50 µm), light of the commonly used wavelengths of 0.8 to 1.55 µm will propagate in several modes, i.e., forming a multimode fiber. A multimode fiber is shown in Fig. 4.9. In a thin core with a diameter of less than 0.85 µm, light propagates in a single mode, as shown in Fig. 4.10. Besides the propagation properties, there are other differences between thin and thick fibers. For instance, multimode fibers are easier to splice and couple, while single-mode fibers have lower loss due to lower dopant concentration in the core. Independent of the thickness of the core, fibers loose light energy due to scattering in the glass and absorption due to dopants and other impurities. Multimode fibers are subject to modal dispersion due to the difference in velocity between the modes. Single-mode fibers do not have this problem and thus are preferred for high-speed transmissions.

12.4 Optical Switch Components

A widely discussed component for an optical switch is a *directional coupler*. It is a 2 × 2 optical switch that directs two incoming light

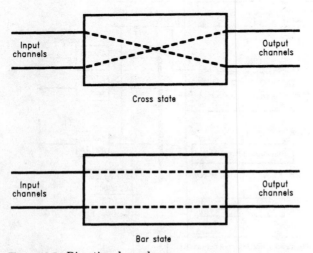

Figure 12.5 Directional coupler.

waves to two outputs, as shown in Fig. 12.5. The two light waves can pass through the directional coupler with the upper input channel connected to the upper output channel and the lower input channel connected to the lower output channel. This is known as the *bar state*. Alternatively, the upper input channel can be connected to the lower output channel and the lower input channel to the upper output channel. This is known as the *cross state*.

Directional couplers are typically fabricated from a titanium-diffused lithium niobate (Ti:LiNbO$_3$) substrate and take the form shown in Fig. 12.6. Two waveguides are formed in the substrate through diffusion of titanium in the shape of two channels. The processes used are similar to those applied in the fabrication of semiconductor devices. In selecting the materials of which a waveguide is formed, it is essential that the index of refraction be higher for the material forming the waveguide than for that of the surrounding material. Thus titanium (Ti) is selected for the waveguides and lithium niobate (LiNbO$_3$) for the surrounding substrate.

The switching principle of directional couplers is based on the fact that the optical energy of one waveguide is coupled to the other waveguide due to the overlap in the evanescent fields of the two waveguides when certain conditions are met. One condition is that the two waveguides have to be close to each other (distance l in Fig. 12.6) for a certain distance (L in the figure). The coupler is switched between the bar and cross stages depending on voltages applied to electrodes positioned above the channel wells, as shown in Fig. 12.7. The electrodes are separated from the waveguide channels by an insulating layer, typically of silicon dioxide (SiO$_2$).

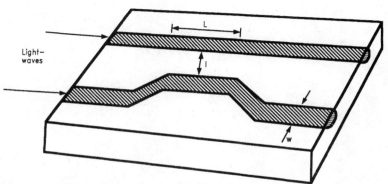

Figure 12.6 Solid-state directional coupler.

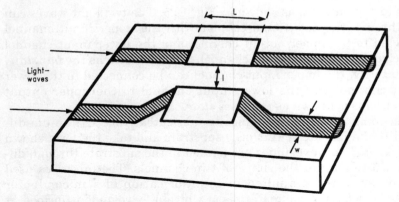

Figure 12.7 Solid-state directional coupler with electrodes.

References

Banniza, Thomas R., Gert J. Eilenberger (SEL Alcatel, Stuttgart), Bart Pauwels, and Yves Therasse (Alcatel Bell Telephone, Antwerp) (1991), "Design and technology aspects of VLSIs for ATM switches," *IEEE Journal on Selected Areas in Communications,* vol. 9, no. 8, October, pp. 1255–1264.

Barri, P., R. Dierckx, J. Goubert, and G. Eilenberger (1990), "Technology for broadband switching," *Electrical Communication,* vol. 64, no. 2/3, pp. 166–176.

Political Aspects

13

International Aspects

13.1 Trends in Foreign Countries

The trends toward broadband networks are basically the same in all industrialized countries. Actually, Europe and Japan are slightly ahead of the United States. The first broadband trial was conducted by the German Bundespost Telekom and German telecommunications manufacturers in Berlin, Germany, between 1986 and 1992 under the name BERKOM. A dozen European countries expect to have commercial international broadband ISDN service up and running in the middle of 1994.

Like the Clinton administration in the United States, many foreign governments promote "electronic superhighways." The difference is that—until recently—foreign governments owned and operated the telecommunications facilities. Thus they control the installation of the required facilities. In addition to the national governments, the European Parliament is urging the telecommunications providers to demonstrate internetworking over cross-border fiberoptic networks.

With the beginning of privatization of telecommunications facilities in Europe, the providers are concerned about competition not only from other European companies but also from non-European firms, in particular U.S. companies. This has lead to accelerated introduction of new services such as broadband in Europe.

13.2 Global Networks

The first international telephone line was opened between Detroit in the United States and Windsor in Canada in 1880. Since then, international telephone lines for voice traffic have grown to reach practically every corner of the earth. International direct distance dialing (IDDD) started in 1970.

Through the use of modems, i.e., modulators/demodulators, to convert digital signals to analog ones, data can be sent on ordinary telephone lines, including international lines. Several companies use dedicated lines to provide international data networks. Many of these networks are combined to form *networks of networks,* such as Internet. Originally, Internet was used by universities, research centers, and government agencies. Today, commercial companies can connect to Internet and send commercial traffic over the network worldwide. The Internet is mainly used to send data and E-mail (electronic mail).

The first steps toward international broadband networks and traffic were taken within CCITT, the international standardization body for telephony and telegraphy. As mentioned earlier, this organization is now known as the Telecommunications Standardization Sector of the International Telecommunications Union (ITU-T). Work on standards for an integrated services digital network (ISDN) started in the late 1970s. Government telecommunications authorities wanted to transmit data and other digital traffic over existing transmission facilities, including subscriber loops. By the end of the 1980s, standards for broadband networks and services were desired, leading to the first steps toward the broadband integrated services digital network (B-ISDN) standard. The idea is a network to which terminals from different vendors can be connected and interwork. Unfortunately, ITU-T has been slow in advancing international standards for B-ISDN and the underlying technique, asynchronous transfer mode (ATM).

In the meantime, the marketplace is not waiting. While ITU-T is taking the top-down approach, the market is taking the bottom-up approach. In other words, broadband is evolving from local area networks (LANs). This is an area that the ITU-T has not approached yet. As a result, manufacturers and users in the United States have come together to form their own standards body, the ATM Forum. This body is working on ATM standards. The support for ATM Forum is expanding to European companies.

Besides manufacturers/users-sponsored standards bodies like the ATM Forum, regional standards bodies have been formed to fill the gap caused by the lack of international standards. Thus the United States has the T1 committee, the Europeans have formed the European Telecommunications Standards Institute (ETSI), and Japan has the Telecommunications Technology Committee (TTC).

In many cases in the history of standardization, countries and regions have been unable to agree on a single standard, to the deprivation of users and customers. The result is that users cannot benefit from low costs based on large-scale production. Examples are the T-carrier systems in North America and Japan compared with the different carrier systems in the rest of the world. The synchronous optical network (SONET) in the United States versus the synchronous digital hierarchy (SDH) everywhere else is another example. Not to mention the metric system. Today, we risk similar problems in broadband technology. The ATM standards being developed by ITU-T, the ATM Forum, and ETSI are not compatible. In other words, products made to one standard do not always work with one of the others. This is in particular true in the user-to-network signaling area. The organizations are trying to remedy this through increased cooperation.

13.3 Broadband ISDN as a Global Concept

When work on broadband ISDN started in CCITT in the late 1980s, the concept was to create a ubiquitous network like the voice telephone network. Not only should it be possible to transmit broadband information on a switched worldwide network, it also should be possible to travel with terminals and connect them anywhere. This is important in order to create a large market with low prices for terminals. This is likely to happen in Europe and the United States. However, it is not necessarily true that the European and U.S. standards will be the same.

Another problem is that the introduction of broadband facilities in the United States will take longer than the introduction of similar facilities in Europe.

14

The FCC, The U.S. Congress, Etc.

14.1 Introduction

Lately, *electronic superhighways* have become a major political item in the United States. In 1992, the Clinton-Gore campaign started promoting information superhighways as a means to improve U.S. competitiveness. The Clinton administration continues to work for such highways for the transportation of medical, educational, and other information and entertainment. In particular, the transportation of images and video is emphasized. Initially, it sounded like the government wanted to build these superhighways. Soon it was realized that the federal budget could not afford it and that private industry was already well under way in designing and building links with high bit rates. Thus, in 1994, the administration started coaxing rather than forcing the telecommunications industry and it started building federal networks.

In a speech in Los Angeles on January 11, 1994, Vice President Gore outlined the Clinton administration's new telecommunications policy. He said that the administration wants competition between service providers, including letting telephone companies and cable television companies enter each other's businesses, at the same time recommending as little regulation as possible but still preventing unrestrained monopolies.

This means changing the Communications Act of 1934 and the "judgments" between AT&T and the U.S. Department of Justice, such as the modified final judgment (MFJ) that split the Bell System in 1984.

We are used to ubiquitous telephone service. This includes offering telephones at affordable prices in rural and other remote areas and to poor people, so-called *universal service*. This service traditionally has been subsidized by long-distance calls and by having artificially high fees for businesses. Today, some of the money still comes from interexchange carriers (IXCs), who provide long-distance service and who pay access fees to the local exchange carriers (LECs) to the tune of $26 billion annually. They in turn use the money to subsidize telephones in remote areas. Recently in the United States, other companies have been allowed to compete with the LECs for local traffic and to bypass them at the same time since they do not have to contribute subsidies to keep all subscriptions affordable. This new competition will make it difficult for LECs to maintain high fees for businesses. So where are the subsidies to come from, the ones needed to maintain low-cost basic service and universal service even in the new multimedia information age? Who will pay for video dial tones in remote areas? This is an expensive proposition, and it can hardly be accomplished at so-called affordable fees without some sort of subsidies.

In his speech, Vice President Gore failed to address details. Industry watchers say that these have to be worked out by the administration (in particular the Department of Justice), Congress, and the Federal Communications Commission (FCC).

14.2 Monopoly versus Competition

A nationwide telegraph or telephone network tends to become a monopoly and requires regulation to maintain reasonable fees and ubiquitous service. Also, once the state grants a unique permit to provide telegraph or telephone service, that permit usually prevents others from competing. In such a situation, only the telephone company or authority has the right to let telephone wires and cables cross a public street.

In the early 1980s, the U.S. Department of Justice decided that the American Telephone and Telegraph Company (AT&T), with close to 80 percent of all telephones, had become too strong a monopoly and forced its divestiture. Thus, since January 1 of 1984, the local telephone service of the former Bell System has been broken up into seven regional telephone companies, the regional Bell operating companies (RBOCs). Even earlier, other companies providing long-distance telephone service in competition with AT&T entered the market. After the divestiture, AT&T remained as the major provider of long-distance services. It should be noted, however, that there are some 500 other companies providing long-distance telephone service in the United States, in addition to about 1300 local operating companies.

Thus we see a peak in the concentration of telephone service in the mid-1980s. This fact is true not only in the United States but also worldwide. Many countries have followed the trend in the United States and allowed competing entities to provide telephone and other telecommunications services.

In the early days of the telephone, you often found multiple companies providing service in the same area. As late as the mid-1940s, there were two local telephone companies covering Mexico City. This meant that professionals and businesses had to have at least one telephone connected to each of the two companies. Eventually, it became possible to place calls between users that were connected to different networks. It was like placing a long-distance call within the same city. We mention this to point out that there is nothing inherent in limiting the provision of telephone service in an area to one single entity. Actually, competing local telephone service has already returned. The FCC, for instance, grants licenses to provide cellular telephone services to two different companies in the same area. Not only that, the possibility of having other companies competing with the local telephone companies in providing telephone service has been granted in some U.S. cities. Companies are allowed to bypass the local companies and connect directly to providers of long-distance services.

There are many pros and cons associated with whether telecommunication services are offered by a monopoly or by competing companies, as shown in Table 14.1. Generally speaking, the installation and operation of a single network is less expensive than having duplicated facilities. On the other hand, competition tends to lead to lower fees for users. A provider that has a monopoly has to be regulated by a public entity in order to ensure ubiquitous service at fair prices. With free competition, there is a risk that the provider will service profitable accounts only, so-called cream skimming. In particular, unprofitable telephone service, such as to remote rural areas, will not be provided or will be offered at exorbitant prices. In a monopolistic situation, there are few incentives to offer new and better services or to

TABLE 14.1   Comparison Between Monopoly and Competition

Monopoly	Competition
Cost-effective	Lower fees
Regulation	Freedom
Ubiquitous service	Cream skimming
Limited types of service	Users' demands met

meet consumer demands. Only through pressure from the regulatory agency, which is likely to be long in coming, will new consumer demands be met.

Not only that, a monopolistic operator, including a state agency, will protect itself from competition. In many countries and for a long time, users were prohibited from sending anything but clear voice messages over telephone lines. You could not send data or facsimile transmissions over the lines. You were not allowed to connect modems to the lines, even through so-called acoustic coupling. With this kind of coupling, the modem was not connected directly to the telephone lines, but it generated sound within the voice band that could pass through the telephone instrument to the network. At the other end, another modem listened to the signals and converted them back to data. Thus there were no physical connection, just an acoustic connection with the receiver and mouthpiece of the telephone set, respectively.

In many cases, the official reason given was that the telephone network could be damaged by "unauthorized" equipment. The unofficial reason, however, was often that the telecommunications authority wanted to protect some other, more profitable service, such as telegraph or telex.

14.3   Telephone Companies versus Cable Television Companies

Both telephone companies and cable television operators have networks of cables that reach most residential users and businesses. Current regulations in the United States mandate that telephone companies can provide telecommunications services only, and cable companies television services only. Today, both categories want to expand into the other group's territory; i.e., telephone companies want to provide television programs, and cable television companies want to provide telecommunications services. Such permits have already been given in a few areas. In preparation for this, we see mergers and acquisitions between telecommunications companies and cable television companies. At one time Bell Atlantic, one of the regional Bell operating companies (RBOCs) negotiated to acquire Tele-Communications, Inc. (TCI), the largest cable television company in the United States. Time Warner, the second largest cable television provider, will provide cable television in the territory of U.S. West, another RBOC, and U.S. West will assist Time Warner in providing telecommunications services outside that territory. Some politicians are concerned that this trend will lead to a few monopolistic providers.

Management

15

Availability, Reliability, and Security of Connections

15.1 Introduction

The availability and reliability of a network should be judged in relation to what the users expect. This is discussed under "Performance Criteria" in Sec. 16.1 of Chap. 16. The main concerns of users are the speed, accuracy, and dependability of such functions as call setup, transfer of user information, and release of connections. A specific concern is the overloading of the network resulting in congestion. Some users are also concerned about the security of connections. Encryption of user information is a means to increase the security.

Networks are subject to internal failures, as well as to interferences from outside persons, specifically so-called hackers. In the 1970s, some people used a so-called blue box to send tone signals to the analog telephone network in order to redirect calls and make long-distance calls for free. With today's and future digital networks, hackers are trying to enter the computers that control the networks and leave so-called viruses, bugs, worms, etc. These can change billing, redirect calls, or just immobilize the network. Steps have to be taken to protect the networks from these kinds of abuses. One reason for private enterprise networks is to protect the users from the abuses and failures that occur on public networks.

Information networks are becoming more and more important for business. Billions of dollars are transferred between banks electronically; just a few minutes of delay can cost millions. The medical profession uses electronic transfer of images and patient information, and here delays can cost lives.

Future ATM-based broadband networks will carry a mixture of voice, data, image, and video traffic. The performance criteria for these kinds of traffic will be different. The type of traffic also will vary from continuous to bursty and from low to high bit rates. Both the average and the peak rates of traffic are measured and considered.

ATM traffic from different sources that generate traffic of a variety of bit rates and burstiness will be concentrated by multiplexors and/or switches. This can be performed in two distinct forms of multiplexing: deterministic or statistical. As an example, Fig. 15.1 shows three channels being multiplexed into one channel. The top channel to the left has an average relative throughput of 4.4. This number could refer to millions of bits or cells, for example. The middle channel has an average throughput of 3.4, and the bottom channel has an average relative throughput of 3.2. All three have a peak relative throughput of 9. The traffic from these three channels is multiplexed onto a single channel with an average relative throughput of 10.9 and a peak relative throughput of 19. This constitutes *statistical multiplexing* because the sum of the peak relative throughput of the three individual channels (27) is more than the relative throughput capacity of the main channel, which is set at a relative throughput of 20. Statistical multiplexing is possible because the sum of the average relative throughputs of the three channels (4.4 + 3.4 + 3.2 = 11) is less than the capacity of the main channel.

In the case of *deterministic multiplexing,* the sum of the peak relative throughputs of the three constituent channels must be less than the peak relative throughput capability of the main channel. In other words, in the example, the capacity of the main channel must be a relative throughput of (9 + 9 + 9 =) 27 or more. The conclusions drawn from this example presume that the relative throughput on any of the channels never is larger than the stated peak relative throughput.

15.2   Congestion Control

15.2.1   Introduction

Congestion occurs when user traffic competes for the services of limited resources in a network. In a circuit-switched network such as the conventional telephone network, congestion on links or in nodes results in a busy signal to the caller. Some networks differentiate between the signal sent when the called party is busy and when links and/or nodes are congested. Most circuit-switched networks apply alternate routing to lead traffic around trouble spots and thus to alleviate congestion.

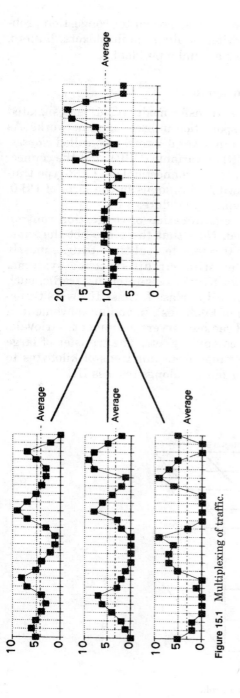

Figure 15.1 Multiplexing of traffic.

Packet- and cell-switched networks also encounter congestion problems. Congestion can occur in switches, links, multiplexors, buffers, etc. Here, too, traffic can be routed around a problem.

15.2.2   Classification of congestion control

As discussed earlier, asynchronous transfer mode (ATM) is designated to become the major form of transportation in broadband networks. As shown in Fig. 3.26, ATM appears in several forms, also called *classes*. The bit rate can be constant (CBR) or variable (VBR), and the connection mode can be connection oriented or connectionless. CBR-type traffic typically comes from circuit-switched systems in the form of DS-0, DS-1, DS-3, E-1, E-2, etc. Video typically produces VBR.

Figure 15.2 shows a common classification of congestion control. Four classes of traffic are defined. Note that these are not the same classes as those of ATM. Class 0 refers to real-time traffic, mostly voice, video, and traffic from existing circuit-switched systems. Classes I, II, and III refer to *data traffic*. Low-intensity traffic, such as interactive data traffic, is grouped in class I. Class II covers transfers of short files (typically tens of kilobytes), including movement of pages between workstations and remote servers, document retrievals, exchange of data in distributed computing, etc. The transfer of large files or images (typically in the range from hundreds of kilobytes to several hundreds of megabytes or more) belongs to class III.

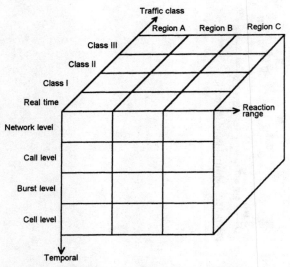

Figure 15.2 Classification of congestion control.

The temporal layers are based on the differences in the time constants between layers of the network. The cell level is the lowest level in the hierarchy and controls the flow of cells in ATM. In the case of bursty traffic, a link can be idle for a time and then suddenly be subject to a burst of traffic. The burst-level flow control controls the traffic flow when bursts occur. The call-level control determines whether a new virtual circuit (VC) with a given bandwidth requirement should be accepted or denied service. The network-level flow control is at the top of the hierarchy and handles the overall control of the network. It determines whether changes in the network are required due to short-term and/or long-term changes in traffic profile and characteristics. For instance, these can change by the time of day or by seasons.

The third dimension classifies congestion control according to the reaction range, i.e., the reaction time inside a network. Such reaction times include the latency in detection and reaction to congestion and the network time constant (NTC) T_N. NTC is a function of link rates, size of buffers, and maximum tolerable delay at nodes. The reaction time or delay in the system is called T_R.

Based on T_N and T_R, three regions of reaction ranges are defined. If $T_R \ll T_N$, then the reaction range is said to be in region A. With $T_R \approx T_N$, we are in region B. Region C is defined for $T_R \gg T_N$.

15.2.3 Congestion control

Congestion control can be either preventive or reactive. *Preventive control* takes place before the fact and tries to prevent congestion. *Reactive control* tries to remedy the consequences of a congestion that has already occurred. With the high transmission speeds of asynchronous transfer mode (ATM), it is difficult to use reactive control. ATM networks have buffers to control overload and/or memories to store information, but these are limited in size. In order to be effective for flow control, the product of the channel bandwidth and the propagation delay of the link must be smaller than the size of the buffers and/or memories. In most cases, they would overflow before any action can be taken.

Several preventive control methods are available. One type of congestion occurs when the network accepts more traffic than it can handle without impairing the quality of service (QOS). Thus admission control is used to limit the amount of traffic entering the network in order to keep the balance between QOS and the utilization of network resources. It takes place at the access points to the network and accepts a new call only if enough network resources are available to maintain the contracted QOS. In order to do this, the admission control scheme needs to know the current traffic flow situation of the network and the characteristics of the flow expected from the new call. Several different schemes to measure current traffic and predict future traffic exist.

The unlimited acceptance of calls that require a large bandwidth may use up bandwidth resources to the extent that smaller calls are blocked. One way of solving this problem is to limit the maximum bandwidth of a new call to some percentage of the available bandwidth. However, this may cause increased blocking of high-bandwidth calls. In the case of frequent calls with large bandwidth between two points, one would consider allocating a fixed virtual path between these points. A drawback of this arrangement is that the allocated bandwidth may remain unused for long periods of time.

Special control may be advisable for bursty traffic from a source that may be inactive for long periods between bursts. Rather than keeping a connection open all the time, the source may be required to renegotiate a new connection before each burst. This may be accomplished by having the source transmit a reservation cell along the virtual channel (VC) before each burst. This cell should contain information about the characteristics of the pending burst. Depending on the available resources along the VC, the transmission of the burst is allowed or denied. In the case of connectionless service, it may be advisable to transmit a burst immediately after the reservation cell without waiting for acknowledgment and to let the cells be discarded if and when there is congestion at some node.

The information flow in a network still has to be controlled after a connection is set up in order to maintain fairness and good performance. One well-known form of policing a network is called the "leaky bucket." Figure 15.3 shows a leaky bucket that accepts a stream of incoming cells and sends out cells to a communications link from its leaky bottom. The size of the bucket corresponds to the maxi-

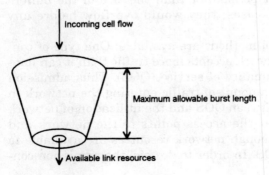

Incoming cell flow

Maximum allowable burst length

Available link resources

Figure 15.3 Leaky bucket.

mum cells permitted in a burst. The overflow of the bucket indicates transmission overflow. Implemented in a network, cells that overflow are discarded. In a virtual leaky bucket concept, overflowing cells are not immediately discarded. Rather, they are marked to be discarded in a congestion situation down the path.

Reactive control can be implemented through the use of the *cell loss priority* (CLP) *bit* in the ATM header. In a congestion situation, cells with a low priority according to the CLP bit are discarded. The CLP bit can be set by the user or by a leaky bucket mechanism.

In other approaches, information about a congestion situation is forwarded along the path to the destination in the form of an *explicit congestion notification* (ECN). The destination then sends back a message to the source to limit its transmission. Another name for this mechanism is *forward congestion indicator* (FCI). Rather than going through the destination, congestion information can be sent from the point of congestion to the source directly through *backward congestion notification* (BCN). Cells also can be rerouted around a congestion.

Distributed source control (DSC) has been suggested by Ramamurthy et al. (1991) specifically for flow control of high-speed data traffic such as file transfers, image retrievals, etc. It is implemented in the ATM layer and controls the rate at which cells can access the network. DSC is based on two parameters: a bandwidth window W_S and a smoothing interval T_S. The control strategy is implemented through negotiation of the two parameters W_S and T_S. The latter should be kept below the *network time constant* (NTC).

An effective flow control in a multimedia environment requires the application of several of the types of control discussed above. It should involve all or most levels in the classification scheme and is thus called *multilevel control*.

15.3 Security

Communications networks are subject to a range of threats. Among them are stealing of equipment and software, unauthorized access to data bases, unauthorized changes of software and hardware, etc. The modern concept of networks is that they should be open and accessible. Obviously, this is in contradiction to keeping them secure. Thus the access has to be limited through the use of passwords, magnetic cards, etc.

None of this is specific for broadband networks. The multimedia environment in which broadband networks are likely to operate involve a diversity of traffic and bit rates. Video-on-demand requires secure means of obtaining payment for the source. Medical images, bank records, money transfers, etc., must be kept secure from unauthorized access and manipulation. With information from many sources mixed on packet- and cell-switched networks, keeping individual data secure poses specific problems.

In particular, national defense entities are concerned about communications security. Starting with the U.S. Department of Defense, several bodies are working on standards for the security of data bases. Among such standards are the Trusted Computer Security Evaluation Criteria (TCSEC) in the United States and the Information Technology Security Evaluation Criteria (ITSEC) in Europe. Both standards specify security aspects of data bases, computers, operating systems, terminals, switches, transmission equipment, etc. Each individual line in a software program has to be evaluated and rated. An entire network or a part of a network is evaluated and given a security rating on a seven-level scale. No commercial system has yet been rated above the fourth level. The rating is specific for a piece of equipment in a specific environment. Thus equipment cannot be moved from one use to another. The evaluation process is very time consuming, and the testing of a component typically takes 12 to 18 months. The rating of an entire system can take 3 years.

15.4 Encryption

Text and signals that are intelligible to the sender, the receiver, and to anybody else are said to be in *clear text* or *plaintext*. *Encryption* is a set of methods by which plaintext is converted into an unintelligible form through the processes of *enciphering* and *encoding*. The distinction between *cipher* and *code* is hazy and lacking a universally accepted definition. One way of distinguishing between them is to state that a code represents units of plaintext, while a unit of cipher (often a single character or bit) represents a single unit of plaintext with a constant length. Plaintext can be transformed into ciphertext through *transposition* or *substitution*. In transposition, the individual characters in the plaintext are maintained in the ciphertext, but their order is changed. In substitution, a specific cipher character or symbol is substituted for each plaintext character. In a *product cipher,* a combination of transposition and substitution is used. A *key* is used to encipher plaintext and decipher ciphertext. It is a set of bits, characters, numbers, or rules that specify such things as the arrangement of characters within a cipheralphabet, the pattern of shuffling in transposition, or the settings of a cipher machine. If a specific key is used for each communication between two parties, the number of keys increases with the square of the number of parties.

This leads to problems with the administration of these keys. A solution to this problem is to use *public keys*. In public-key cryptography, two keys are used, one private and one public. As the name implies, the public key is made available to anyone, while the private key is kept secure. The private key can be used either for encryption or decryption, while the public key is used for the opposite operation. The concept of public keys was introduced by Whitfield Diffie and Martin Hellman of Stanford University in 1976. R. Rivest, A. Shamir, and L. Adleman (RSA), also at Stanford University, came up with a well-known public-key system in 1978. It is based on two randomly selected prime numbers P and Q from which the product N is calculated. A public exponent e is also chosen. The values of N and e form the public key, while the two prime numbers form the private key. In order to encipher a message M, M^e mod N is calculated. Deciphering involves calculating $(M^e)^d = M$, where d can be obtained by anyone knowing the factors of N. *Cryptanalysis* refers to the process of searching for and finding the plaintext without the benefit of the key to the ciphertext.

When substitution is used on the character level, it is fairly easy to decipher the ciphertext through the use of *frequency analysis*. In any plaintext, characters and groups of characters appear according to known frequencies. Thus, by establishing the frequency of characters in a ciphertext, the corresponding plaintext characters can be identified. Because the character-frequency tables are specific for every language, it is also possible to first determine in what language the ciphertext is written.

The more ciphertext a cryptanalyst has available, the easier it is to decipher the material. In order to foil the analysts, encryption keys should be changed often, even several times in the same message. The most secure encryption is made by using a key of the same length as the plaintext and using that key once only.

Enciphering of data is commonly done on the bit level, but it also can be done on the byte or higher level. Typically, a bit is changed from a one to a zero, or vice versa, or kept unchanged according to some system and a key. One such encryption system is the "U.S. Data Encryption Standard" (DES), which was published in 1977. It uses a product cipher and operates on a block of 64 bits using a 64-bit key. Systems such as DES are called *symmetric*, contrary to public-key systems.

The security of the DES was questioned already at the time it was introduced. Martin E. Hellman and Whitfield Diffie of Stanford University wrote in *Computer* of June 1977 that a special machine for the unique purpose of cryptanalyzing DES ciphertext could be built for $20 million (in 1977) and obtain the secret key in 12 hours.

Responding to the attacks on the DES, statements in the U.S. Congress and by federal authorities indicated that the DES with a 56-bit key is secure for "at least 10 years." Those 10 years ran out in the late 1980s.

The U.S. National Security Agency (NSA) is promoting a new standard for voice communications called "Clipper" and another for data communications called "Tessera." As a crime-prevention feature, both standards have trapdoors allowing the government's law enforcement agencies to decipher the messages. This has met with opposition from civil liberties groups. Further, the NSA will not release the architecture of the cipher standards, thus preventing private specialists from evaluating them.

Private industry is promoting a competing commercial encryption standard, called the *RSA Public Key Cryptosystem*. It is developed by RSA Data Security, a Redwood City, California, company, and is based on the above-mentioned algorithm developed by Rivest, Shamir, and Adleman. The system is endorsed by leading computer and telecommunications companies.

The most secure form of encryption is when the sender personally encrypts the plaintext and the receiver personally decrypts the ciphertext. Alternatively, encryption devices can be used at different locations in a network, as shown in Fig. 15.4. Again, the closer the encryption is done to the sender and the decryption to the receiver, the better it is from the security point of view. Whenever a message travels in plaintext, it is subject to threats from the "enemy." Thus encryption/decryption devices should be placed inside personal computers or attached directly to their communications ports in such a way that unauthorized persons cannot access them. Figure 15.4 also shows a case where the encryption/decryption device is placed in or at the trunk side of a private branch exchange (PBX). The reason for this may be to save expenses by having many people share the same device. Radio communications links such as microwave and satellite links are particularly subject to threats. It is very easy for the "enemy" (whether commercial or military) to listen to and record messages sent over the air. Thus such links are often protected by encrypting the traffic over the link, as shown at the top of Fig. 15.4.

Today's digital communications lend themselves to direct encryption on the bit level. The high bit rates of broadband communications make it difficult to encrypt and decrypt in real time. Both the encryption and the decryption take time (even though limited) and add delays to the transmission. Thus voice and video in particular can be deteriorated. However, the Sandia National Laboratories encrypted a DS-3 link between Albuquerque, New Mexico, and Livermore,

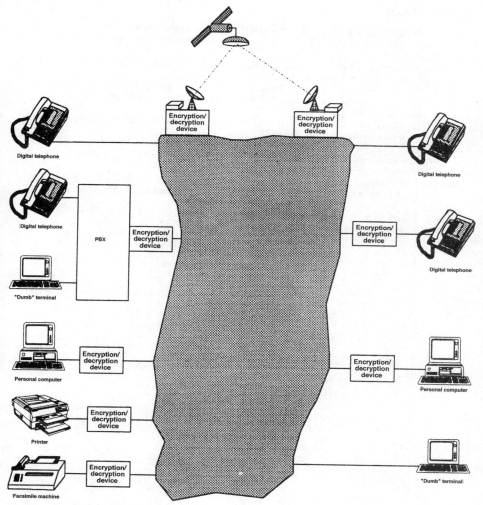

Figure 15.4 Potential locations of encryption devices in a network

California, using a pair of KG95 encryptors and found the delay to be zero and the error rate to be less than 10^{-13} (Naegle, 1994).

Care must be taken with commands that must be available during transmission. They cannot be encrypted. This includes headers and tails in packets and cells. Thus it is preferred to encrypt the messages at the source and then embed them in the common transmission protocols during transmission, retrieve the encrypted message after transmission, and decrypt it.

References

Ford, Warwick, and Brian O'Higgins (1992), "Public-key cryptography and open systems interconnection," *IEEE Communications Magazine,* vol. 30, no. 7, July, pp. 30–35.

Fraser, Alexander G. (1991), "Designing a public data network," *IEEE Communications Magazine,* vol. 29, no. 10, October, pp. 31–35.

Habib, Ibrahim W., and Tarek N. Saadawi (1991), "Controlling flow and avoiding congestion in broadband networks," *IEEE Communications Magazine,* vol. 29, no. 10, October, pp. 46–53.

Lindberg, Bertil (1981), *Data, Text and Voice Encryption Equipment,* International Resource Development, Inc., Norwalk, Conn.

Naegle, John H., Nicholas Testi, and Steven A. Gossage (1994), "Developing an ATM network at Sandia National Laboratories," *Telecommunications* (Americas edition), vol. 28, no. 2, February, pp. 21–22, 24.

Ramamurthy, Gopalakrishnan, and Rajiv S. Dighe (1991), "A multidimensional framework for congestion control in B-ISDN," *IEEE Journal on Selected Areas in Communications,* vol. 9, no. 9, December, pp. 1440–1451.

Yazid, Setiadi, and H. T. Mouth (1992), "Congestion control methods for BISDN," *IEEE Communications Magazine,* vol. 30, no. 7, July, pp. 42–47.

16

Network Management Policies

16.1 Performance Criteria

Performance criteria are well established for conventional voice traffic. During the 1950s and 1960s, considerable work was done to ascertain user preferences for voice connections under various conditions and interferences. Subjective tests were conducted where test subjects were asked to rate a connection as "good or better," "poor or unsatisfactory," etc. The purpose of these tests was to obtain the quality perceived by users. Most of the development work was conducted by AT&T and became the standard for the Bell System, which at the time handled about 80 percent of the telephone service in the United States.

Lately, similar performance criteria have been established for today's multiservice environment. While noise can be measured in yesterday's analog systems, it is difficult to measure the quality of information transmitted in digital form. Bit-error rates (BERs) and error-free seconds (EFSs) can be measured easily, but what does it mean to the quality of the information carried?

If bits or packets are lost or corrupted in the transmission of data files, the lost information can and has to be retransmitted. A missing bit in a digital facsimile transmission can corrupt the rest of the facsimile. Again, retransmission is the solution. This is not possible with voice and video, and degradation of the transmitted information will occur.

Information will be transmitted at a range of bit rates, from full uncompressed rates to highly compressed rates. In the case of voice, bit rates will range from over 64 kbit/s to a few kilobits per second. Loss or corruption of bits in highly compressed data will cause more distortion than loss at the uncompressed rate. The same is true for video transmissions, which will range from high-definition television (HDTV) to video compressed to 64 kbit/s.

New transmission technologies and new types of information will cause new types of distortion. The asynchronous transfer mode (ATM) to be used for broadband ISDN is subject to new specific impairments such as cell loss and cell delay variation. These are not present in synchronous transmissions. Errors in audio transmissions may cause phonemic distortions rather than the conventional "pops and clicks." Video transmissions may be subject to streaks. Variations in temporal delays may cause warping if delays in buffers are not handled advantageously. This will be seen as jerky motions.

In an integrated multimedia system, it will be desirable to have a single standard for measuring quality, rather than separate standards for data, voice, images, and video. Cross-disturbances between the media also should be taken into consideration.

In some applications, users have the option to trade spatial and temporal performance aspects. For example, in some videophone systems, a clear picture at a low frame rate can be traded against a blurry picture at a high frame rate. A user can select the spatial-temporal setting based on personal preference or type of application.

Performance criteria for data communications were first incorporated in the American National Standards Institute (ANSI) Standard X.3.102-1983. Corresponding measurement methods are covered in ANSI Standard X3.141-1987. Early international standards covering performance criteria in connection with data communications and ISDN are included in CCITT Recommendations X.140 and I.350, respectively.

The developers of these standards used a criteria/function matrix of the type shown in Fig. 16.1. It shows the basic functions of a circuit switched call/connection along the vertical axis, i.e., setting up the call, transferring user information, and releasing the connection. The performance criteria considered important to most users include speed, accuracy, and dependability, as listed along the horizontal axis. The cells in the matrix identify pairs of functions and criteria that were considered individually by the developers of the standards. The "blocks" mentioned in some cells refer to blocks of user data. The text in the matrix cells describes how users perceive the limitations and/or failures of the data transport system. The speed criterion of the access function is perceived as the time required to set up a call or a connection. Users measure the speed of transferring user information as the time required to transfer a block of data or as the transfer bit rate for user information. The time it takes to release a connection gives the speed criterion for that function.

Criterion / Function	Speed	Accuracy	Dependability
Call setup	Call setup time	Probability of incorrect access	Probability of access denial
Transfer of user information	Block transfer time User information transfer bitrate	Block error probability Probability of misdelivery of user information	Probability of loss of user information
Release of connection	Release time	Probability of failure to release connection	

Figure 16.1 Performance criteria/funtions matrix.

The criteria of accuracy can be expressed as the probability of incorrect access for the access function, the probabilities of block error or misdelivery of user information for the transfer function, and the probability of failure to release the connection for the release function. The dependability criterion for call setup is expressed by its reversed value: the probability of access denial. Similarly, the dependability of the transfer of user information is expressed by another reversed value: the probability of loss of user information. The dependability and accuracy criteria of the release function are listed as identical.

With a little stretching, the criteria/function matrix for circuit-switched networks also can be applied to packet-switched networks, including ATM, as shown in Fig. 16.2. The text "residual error ratio"

Criterion / Function	Speed	Accuracy	Dependability
Call setup	Call setup delay	Probability of call setup error	Probability of call setup failure
Transfer of user information	Transfer delay of data packet Throughput capacity	Residual error ratio	Probability of setup failure Probability of reset Probability of premature disconnect
Release of connection	Delay of "clear" indication	Probability of failure to release connection	

Figure 16.2 Performance criteria/functions matrix for packet switching.

refers to a combination of data error, misdelivery, and loss probability. Two speed criteria for the user information transfer function are listed: transfer delay and throughput. Three dependability criteria are shown for the information transfer function: the probability of setup failure, reset, and premature disconnect. The meaning of the descriptions in the other cells should be self-explanatory.

16.2   Quality of Service

A user negotiates and contracts with the service provider regarding the quality of service (QOS) desired and available. The QOS can be included in the contract that establishes the user's access to the network. It also can be negotiated before (or even during) each connection to the network.

In the case of broadband ISDN or ATM services, the main concerns with respect to the QOS are cell delays and cell loss probabilities. Each application in a multimedia environment is different and may require a different QOS contract. Voice and video traffic, as well as all traffic originating in circuit switched networks, is real-time traffic that cannot tolerate much delay, delay jitter, and cell loss. Data traffic is mainly bursty with long periods without activity. The required QOS for data traffic ranges from relatively low throughput with low delays (latency) to high throughput with moderate delays (latency). Data traffic is sensitive to cell loss but can tolerate some as long as the lost cells are retransmitted later. In the case of voice and video traffic, retransmission of lost cells is impossible due to the delays encountered in this procedure.

16.3   Management

16.3.1   Introduction

Management of a communications network involves many tasks, including those shown in Table 16.1. Management functions can be centralized, distributed, or conducted by users. They can be performed manually or by automatic equipment.

TABLE 16.1   Managing Tasks

Configuration management
Fault detection and isolation
Maintenance tracking
Performance measurement
Applications management
Security
Inventory and accounting

Management of the configuration of a network involves rerouting of traffic depending on changes in traffic patterns and in the availability of resources. Packet and cell switched systems automatically route packets and cells over available and functioning links, avoiding compromised ones. Rerouting is mostly done to alleviate problems due to failed transmission links, switches, and other equipment, including those caused by natural and manmade disasters. Most network transmission equipment has built-in protection switching; i.e., a working channel is automatically substituted for a failed one. When automatic rerouting does not work, manual routing is a solution. Some failures may be so marginal that the automatic equipment does not recognize them. In this case and in the case of overloading due to heavy traffic, manual rerouting is required.

Communications networks can fail for many different reasons. It is part of network management to detect faults and to isolate them, i.e., to determine their location in the network. Table 16.2 lists common network faults and failures.

The most common failure is broken wires, cables, and fibers, typically caused by people. Grounding of wires or shorts between wires, often caused by water, are also common. Fibers in optical cables can be misaligned in splices or impaired in sharp turns. Electromagnetic interference in the form of noise and distortion typically enters electrical wires and cables. This is true especially if the circuits are unbalanced and/or unshielded. The result is distortion of the electrical pulses that carry digital information.

TABLE 16.2   Common Network Faults and Failures

On physical wires, cables, and fibers
  Breaks
  Grounds or shorts
  Water damage
  Misalignment of fibers
  Electromagnetic interference (noise) and distortion

In analog electronic equipment
  Failed components
  Power failure
  Electromagnetic interference (noise) and distortion
  Unbalanced hybrid

In digital equipment
  Failed components
  Power failure
  Impulse and phase distortion
  Malfunctioning sampling gate
  Mismatch in synchronization

Manual tampering

The infrastructure that carries digital information contains analog electronic equipment such as amplifiers, regenerators, multiplexors, modems, and codecs. Even though today's electronic equipment is reliable, failures occur due to failed components, power failure, electromagnetic interference, and unbalanced hybrids. A *hybrid* is a device that interfaces a two-wire, two-way circuit with a four-wire line.

Modern digital equipment is based on very large scale integrated (VLSI) circuits. These are inherently reliable and seldom fail. Still, they do fail and with them the system. Other common failure modes consist of power failure, as well as impulse and phase distortion. Analog signals are sampled in order to convert them to digital signals. The device controlling this, a sampling gate, is subject to failures. Transmission facilities using the synchronous transfer mode (STM) have to be synchronized throughout the network. If the synchronization fails, the pulse signals that carry digital information will be misread.

Finally, people can cause failures intentionally or unintentionally. Manual tampering with equipment is a factor. The most common tests and measurements on communications facilities are listed in Table 16.3.

Lines that carry digital information (in the form of pulses) are basically analog. Their underlying analog features must meet certain criteria before digital signals can be transmitted on them unimpaired. Digital pulses are sensitive to distortion that the human ear ignores. Frequency shift and phase distortion are among them. A common technique for measuring the suitability of a line for digital transmission is called the peak-to-average ratio (PA/R). A perfect pulse train is sent through the circuit under test and the pulse train received at the far end is analyzed. Distortion during the transmission will spread the signal amplitude in time; this spread is measured in the PA/R.

The most common test on digital circuits is to measure the bit error rate (BER) or block error rate (BLER). This test counts the number of

---

TABLE 16.3   Tests and Measurements

Analog tests and measurements on data lines
    test of frequency shift
    test of phase distortion
    PA/R measurements
    test of phase jitter
Digital tests on digital circuits
    test of bit error rate (BER)
    detection of bipolar violations
    redundancy check
    check on synchronization

bit or block errors in a transmission. Some transmission coding schemes use pulse trains with alternating positive and negative pulses. If two consecutive positive (or negative) pulses are received, a bipolar violation is detected, indicating a fault. Bits that do not carry any information added to a bit stream can be used for a so-called redundancy check. It can be used to detect and even correct mistakes. As mentioned earlier, STM requires synchronization, and the network's adherence to synchronization should be tested in STM systems.

The term *maintenance tracking* refers to the bookkeeping of faults and failures in a data base. This information can then be used for statistical analysis of the reliability of equipment, etc.

Performance criteria were discussed earlier. Among the responsibilities of a network manager is to measure the network's adherence to the established criteria, in particular response time and availability.

Application management refers to the arrangement (and rearrangement) of a network to meet the demands of the applications running on it. This includes rerouting of traffic between voice and data depending of the time of day, etc.

A network manager is also responsible for the security of the network, which was discussed earlier. She or he also should keep an inventory of installed equipment and keep accounting records of the network and its utilization.

16.3.2 Central management

Traditionally, management of communications networks has been centralized. Providers of public network services typically have a big screen showing the network and current alarms. This gives the manager(s) an overview of the situation of the network. Reactions to the alarms are typically not taken at the center, but locally, where the alarm originates.

16.3.3 Diversified management

The current trend is to diversify the management of networks and give local managers the possibility and equipment to react to failures, reroute traffic, and accommodate users. Distributed data bases give local managers access to network information without taxing transmission lines and with smaller delays.

16.3.4 Management by users

A further trend is toward letting users manage their access to the network and the allocation of resources for their needs. We have mentioned that users negotiate with network providers and/or the network per se in order to establish performance criteria and request that resources be allocated. The next step is to automate the users' negotiations with the network. On the simplest plane, this means that a user dials the number of the called party to set up a connection. The same procedure is already available for data connections and will soon apply to image and video connections. However, users want to be able to request a specific amount of bandwidth and quality of service. Such a user also will want to indicate the type of traffic (such as continuous or bursty), the priority level, and the expected call duration. Maybe the users also will some day be able to negotiate the price for the connection.

16.3.5 Telecommunication management network (TMN)

The International Telecommunications Union's Telecommunications Sector (ITU-T) is working on an international standard for a telecommunications management network (TMN). This project was started by CCITT as draft Recommendation M.3010. Using signaling system number 7 (SS7), management information will be transmitted between units of the future worldwide broadband networks, such as the B-ISDN.

References

Standards

ANSI Standard X3.102-1983 (1983), "American national standard for information systems—Data communication systems and services—User-oriented performance parameters," ANSI, Inc., New York.

ANSI Standard X3.141-1987 (1987), "American national standard for information systems—Data communication systems and services—Measurement methods for user-oriented performance parameters," ANSI, Inc., New York.

CCITT (1988), "General aspects of quality of service and network performance on digital networks, including ISDN," Recommendation I.350, *CCITT Blue Book,* vol. III, fascicle III.8 (plenary assembly of 1988, published 1989), CCITT, Geneva.

CCITT (1988), "General quality of service parameters for communication via public data networks," Recommendation X.140, *CCITT Blue Book,* vol. VIII, fascicle VIII.3 (plenary assembly of 1988, published 1989), CCITT, Geneva.

Other references

Lindberg, Bertil (1989), *Troubleshooting Communications Facilities—Measurements and Tests on Data and Telecommunications Circuits, Equipment, and Systems,* John Wiley and Sons, Inc., New York.

Seitz, Neal B., Stephen Wolf, Stephen Voran, and Randy Bloomfield (1994), "User-oriented measures of telecommunication quality," *IEEE Communications Magazine,* vol. 32, no. 1, January, pp. 56–66.

Slone, John P., and Ann Drinan (eds.) (1991), *Handbook of Local Area Networks,* Auerbach Publications, Boston.

Wolf, Stephen, C. A. Dvorak, Robert F. Kubichek et al. (1991), "How will we rate telecommunications system performance?" *IEEE Communications Magazine,* vol. 29, no. 10, October, pp. 23–29.

Yazid, Setiadi, and H. T. Mouftah (1992), "Congestion control methods for BISDN," *IEEE Communications Magazine,* vol. 30, no. 7, July, pp. 42–47.

Index

ABOUT THE AUTHOR

Bertil C. Lindberg is an independent consultant, educator, and entrepreneur, based in New York. He has directed and conducted market and technology trend studies in electronics, telecommunications, and data processing worldwide. A respected author in the field of telecommunications, Mr. Lindberg has written and directed many industry studies for publishing houses like Datapro Research Corporation, Frost and Sullivan, and International Resource Development. He is the author of *Troubleshooting Communications Facilities: Measurements and Tests on Data* and *Telecommunications Circuits, Equipment, and Systems*. He has taught at New York University, The City University of New York, and is currently an adjunct associate professor at Brooklyn Polytechnic University.